INTRODUCTORY OF COORDINATION CHEMISTRY

INTRODUCTORY OF COORDINATION CHEMISTRY

Rajesh Bajaj

CYBER TECH PUBLICATIONS
4264/3, Ansari Road, Daryaganj, New Delhi-110002 (India)
Ph.: 011-23244078, 011-43559448 Fax: 011-23280028
E-Mail: cyberpublicationsdelhi@yahoo.com
Website: www.cybertechpublications.com

INTRODUCTORY OF COORDINATION CHEMISTRY

Rajesh Bajaj

First Edition 2010

Published by : Asian Offset Printers
G.S. Rawat for **Cyber Tech Publications**
4264/3, Ansari Road, Daryaganj, New Delhi-110002 (India)
Ph.: 011-23244078, 011-43559448 Fax: 011-23280028
E-Mail: cyberpublicationsdelhi@yahoo.com
Website: www.cybertechpublications.com

PREFACE

The "Introductory of Coordination Chemistry" book is provides all necessary information. While this was gratifying in one sense, at the same time there were many imitating features like typographical errors.

To familiarize the student about expected type of questions in the examinations, important summary and review questions have been provided at the end of each chapter. The subject matter has been presented in a simple and systematic manner. The present edition has many new features. This approach is adopted for some main reasons. Several sections in all these chapters have been rewritten and enlarged to make the text clear. So, the students written some querries and mistakes in this book.

The author shall be gratefully to the readers who point out errors and omissions which inspite of all care might have been there.

— *Author*

Contents

1

Introduction to Complex Compounds

INTRODUCTION

In fact no one method of classification has been found to be completely satisfactory. However, it is at least very instructive to examine some possible methods of classification as these will lead to a deeper understanding of the nature and scope of the coordination compounds. Because of the great variety of compounds that are properly considered as coordination compounds, a classification is difficult, even for study purposes.

Blitz Classification of Complexes

The classification of the complexes on the basis of their stability in solution into the following two classes :

Penetration Complexes

These are the coordination compounds which have sufficient stabilities to retain their identity in solution, *i.e.,* they are not reversibly dissociated in solution like normal complexes $[FeCN)_6]^{4-}$ $[Cu(CN)_4]^{3-}$ and $[Co(NH_3)_6]^{3+}$ are examples of penetration complexes.

$$[Fe(CN)_6]^{4-} \longrightarrow Fe^{2+} + 6CN^-$$

$$[Cu(CN)_4]^{3-} \longrightarrow Cu^+ + 4CN^-$$

$$[Co(NH_3)_6]^{3+} \longrightarrow Co^{3+} + 6NH_3$$

Thus, the ions like $[Fe(CN)_6]^{4-}$, $[Cu(CN)_4]^{3+}$ and $[Co(NH_3)_6]^{3+}$ are penetration complexes because these can be detected as such and there is hardly any evidence of the existence of free Fe^{2+}, Cu^+ and Co^{3+} ions respectively.

The penetration complexes are characterised by a short bond distance between the central ion and donor groups, deep-seated electronic arrangements and are not readily and reversibly dissociated either in the solid or in solution state.

Sometimes the penetration complexes are also referred to as *covalent complexes.*

Blitz recognised that most coordination compounds lie in between normal and penetration complexes. Blitz's classification is of more convenience than of any fundamental importance.

Normal Complexes

These are such complexes which are reversibly dissociated in solution into their constituent species. For examples, $[Cd(CN)_4]^{2-}$ and $[Co(NH_3)_6]^{2+}$ are the complexes of this type.

$$[Cd\ (CN)_4]^{2-} \rightleftharpoons Cd^{2+} + 4CN^-$$

$$[Co(NH_3)_6]^{2+} \rightleftharpoons CO^{2+} + 6NH_3$$

Thus, the complex ions such as $[Cd(CN)_4]^{2-}$ and $[Co(NH_3)_6]^{2+}$ constitute normal complexes because in solution sufficient Cd^{2+} and Co^{2+} ions will exist and can be detected with suitable reagents and tests.

The normal complexes are characterised by relatively weak bonds between the central atom and the donor groups. Magnetic susceptibility measurements of normal complexes reveal that these complexes do not have any deep-seated electronic arrangements.

Sometimes the normal complexes are also referred to as *ionic complexes.*

Second Method of Classification

The coordination compounds may be divided into two groups:

Perfect Complexes

These are those coordination compounds which retain their complex character in solid as well as in solution state. A numberless complex compounds such as K_4 [Fe $(CN)_6$], $[Co(NH_3)_6]$ Cl_2, $[Cu(NH_3)_4SO_4]$, $K_3[Fe(CN)_6]$, etc., fall under this class of coordination compounds.

Imperfect Complexes

Imperfect complex compounds are those co-ordination compounds which remain as complexes either in solution state but not in the solid phase or which exist as complexes in the solid state but break up when dissolved in the solvent.

The imperfect complexes which exist only in solution are K_2 $[Cd(CN)_4]$, $[Cu(NH_3)_2]Cl$, $K_2[CuCl_4]$, $K_2[Ni(CN)_4]$, $[Al\ (H_2O)_6]^{3+}$, etc.

The examples of imperfect complexes which exist only in the solid state are K_2CoCl_4, $Cu_2Cl_2.2CO$, $(NH_4)_2$ $FeCl_4$ etc.

Third Method of Classification

A more general, precise and convincing classification, has also been given.

(a) Type I : Complex compounds belonging to this class are those compounds which contain complex cations or are formed by the union of metal ions (cations) with inorganic molecules such as H_2O and NH_3.

Complexes which contain such complex cations as the ammoniates are $[Zn(NH_3)_4]^{2+}$, $[Cu(NH_3)_6]^{2+}$, $[Ni\ (NH_3)_6]^{2+}$, $[Cd\ (NH_3)_4]^{2+}$, $[Ag\ (NH_3)_2]^+$ and hydrated complex ions like $[Be(H_2O)4]^{2+}$, $[Cr(H_2O)_6]^{3+}$, $[Al\ (H_2O)_6]^{3+}$ etc. The extent to which an ammoniated or hydrated complex is formed with a cation has been found to depend upon the following two factors :

(i) The concentration of ammonia, and

(ii) The stability of the resulting complex.

For instance, Fe^{3+} and Al^{3+} ions, when treated with ammonia, are always precipitated as hydroxides because their ammonia complexes are unstable.

(b) Type II : This class of complex compounds includes those type of complexes which contain only complex anions and are obtained by the combinations of cations with inorganic anions in such a way that the number of anions is invariably greater than the number of anions required to satisfy the electrovalence of the cation.

The familiar examples are the halide complexes like K_2HgI_4, K_2PtCl_6, $(NH_4)_2SnCl_6$,$NaCuCl_2$, $(NH_4)_2PbCl_6$, the cyanide complexes $K_2Cd(CN)_4$, $Na[Ag(CN)_2]$, $K_2[Ni(CN)_4]$ thiocyanate complexes $K[Ag\ (CNS)_2]$,

$K_2[Hg(CNS)_4]$, sulphide complexes such as $(NH_4)_2$ $[AsS_3]$, $(NH_4)_2SnS_3$ etc.

(c) Type III : The co-ordination compounds belonging to this type are made up of a complex cation and complex anion. This complex cation is obtained by the combination of the organic or inorganic molecules or both with the metal ion. Examples representing the class are such compounds as $[Cr(NH_3)_6]$ $[Cr(NSC)_6]$, $[Co(NH_3)_6]$ $[Cr(NSC)_6]$, $[Pt(NH_3)_4]$ $[Pt(Cl_4]$, etc.

Fourth Method of Classification

A fourth method by which one may classify complexes is according to the electronic configuration of the metal atom or ion in question.

Category I :

This includes complexes of metal atoms which have a valence shell with pseudo-inert gas configuration, *i.e.,* (n – 1) d^{10} where *n* is 4, 5 or 6. These central atoms are also spherically symmetrical species and are including some metals in negative oxidation states. All complexes formed by these species (shown in below) are highly covalent.

The big difference between the categories I and II is that for the species in the first category, the *dπ* acceptor orbitals are of lower principal quantum number than the σ-bonding s and *p* orbitals, whereas in this second category the π acceptor orbitals are outer *d* orbitals. The stereochemistry of the complexes formed by these metal atoms is also explained by VSEPR theory and all complexes are diamagnetic.

Mn	Fe	Co	Ni	Cu	Zn	Ga	Ge	As
	Ru	Rh	Pd	Ag	Cd	In	Sn	Sb
	Os	Ir	Pt	Au	Hg	Tl	Pb	Bi

Category II

This includes the complexes of such metal atoms which have pseudo-inert gas plus two configuration, *i.e.,* (n – 1) d^{10}, ns^2, where n is 4 5 or 6. These complexes possess certain geometries (shown in below).

1 +	2+	3+	4+	5+	
Ga	Ge	As	Se	Br	
In	Sn	Sb	Te	I	He
Tl	Pb	Bi	Po	At	Rn

For example : (Se,Te) X_4 compounds have geometries based upon the lone pair occupying a stereochemical site, and the same is true of compounds of Br(V), I(V), Xe(VI), etc., even though these are not generally considered to be central metal atoms.

Category III

This includes complexes of metal atoms which possess incompletely filled *d* orbital, (n – l) d^1 to ^{9} where n is 4, 5 or 6. This group of central atoms (shown in below) is by far the largest and most diverse since it includes all of the transition metals in all of their many oxidation states except those which would place them in categories I and II. The complexes have perfectly regular structures predicted by VSEPR theory.

Ti	V	Cr	Mn	Fe	Co	Ni	Cu
Zr	Nb	Mo	Tc	Ru	Rh	Pd	AE
Hf	Ta	W	Re	Os	Fr	Pt	Au
Th	Pa	U					

Final Remark on Classification : After examining the foregoing four methods by which we might classify complexes, we see that no one method stands out clearly as best and none of them is totally satisfactory.

However, the mere attempt to find a suitable classification system has hopefully led the reader to a greater appreciation of the broad scope of the field and many facets of it to be explored.

POLYDENTATE LIGANDS

Polydentate ligands are said to have flexidentate character if they do not use all its donor atoms to get coordinated to the metal ion. An interesting example of such polydentate ligands is ethylene diamine tetraacetic acid. This ligand generally acts as a *hexodentate* ligand but it acts as a *pentadentate* ligand [*e.g.*, Cr^{III} (OH) (HEDTA)$]^{2-}$ [Co^{III} (Br) (HEDTA)$]^{2-}$ and as a *tetradentate ligand* [*e.g.*, Pd^{II} (H_2EDTA)]°. Another polydentate ligand having flexidentate character is sulphate group which acts as a *monodentate* [*e.g.* Co $(NH_3)_5$ $SO_4]^+$ and *bidentate* [*e.g.*, Co^{III} $(NH_3)_5$ $SO_4]^+$ respectively. This has been confirmed by infrared spectroscopy.

When the infrared spectrum of (Co^{III} $(NH_3)_5$ $SO_4]^+$ is recorded, it shows six separate absorption bonds due to S–O vibrations. This reveals that an oxygen atom of the sulphate group gets covalentaly bonded to Co^{3+}.

When the infrared spectrum of $[Co^{III}(en)_2SO_4]^+$ is recorded, it shows eight bands due to S–O vibrations. This reveals that the sulphate group acts as a bidentate group.

In certain cases ambidentate ligands result in linkage isomerism (for definition, see isomerism).

The complex $[Co(NH_3)_5NO]Cl_2$ has been found to exist in two linkage isomers, *i.e.,* $[Co(NH_3)_5ONO]Cl_2$ (nitrito isomer) and $[Co(NH_3)_5NO_2]Cl_2$ (nitro isomer).

Table 1

Ambidentate ligands	*Metal- ligand bond*
$S_2O_3^{2-}$ ion	Thiosulphato–S(MSSQ$_3$) Thiosulphato–O(MOSO$_2$ S)
R_2SO ion	S–bonded, O–bonded
$SeCN^-$ ion	MSeCN,MNCSe
NCO^- ion	MOCN.MNCO
NCS^- ion	Thiocyanato, MSCN Isothiocyanato, MNCS
CN^- ion	Cyano(MCN) Isocyano (MNC)
NO_{2} – ion	Nitro (M–NO$_2$) Nitrito (M – O–N = O)

KINDS OF THE LIGAND ATOMS

On the basis of formation of complexes with different atoms, Pearson classified ligands as well as the metals into hard and soft ones, *i.e.,*

(a) The metal ions with almost empty or completely filled d- subshell cannot be used for the formation of π bond. These metal ions are called *the hard acids* or *class (a) metals.* Examples of these are alkali metals, alkaline earth metals, Al, Ga, In, Sn, Pb, Ti, Zr, Hf, Y.Sc, La, V, etc. The ligands that form stronger complexes with metals or hard acids are called the *hard bases* or class (a) *bases.*

(b) The metal ions with nearly filled d-subshell can form π bonds with ligands which can accept these *d* orbitals electrons in their empty *d* orbitals or suitable π-orbitals (vacant).

These metal ions are called the *soft acids* or *class* (*b*) *metals* while the ligands are called *soft bases or class* (*b*) *bases.* Examples of soft acids are Cu (I), Hg (II), Pd (II), Pt (II), PR_3. The charge on the ligand is important because it determines many properties of the complex. If the sum of the negative charges of the ligands is equal to the charge on the metal ion, the complex is a *non-electrolyte* otherwise it is an *electrolyte.*

Coordination Number

It may be defined as the *maximum number of groups* (*neutral molecules or negative ions*) *which can he arranged or coordinated around a central metal ion.*

It is usually abbreviated as C.N. In $K_3Fe(CN)_6$. six cyanide ligands are coordinated to Fe^{3+} and hence coordination number of Fe^{3+} is 6. Similarly, in $[Pt(NH_3)_4Cl_2]$ coordination number of Pt^{2+} is 4 and in $[Co(NH_3)_6]Cl_3$ coordination number of Co^{3+} is six.

Coordination number of metal varies from 2 to 10, but the most common coordination numbers are 4 and 6, but may be 2 or 8 or an odd number in rare cases.

Table 2

Metal ion	*C.N.*	*Metal ion*	*C.N.*	*Metal ion*	*C.N.*
Ag^+	2	CU^{2+}	4, 6	OS^{3+}	6
Au^+	2,4	Zn^+	4	Ir^{3+}	6
TI^+	2	Pb^{2+}	4	Au^{3+}	4
Cu^+	2,4	Pt^{2+}	4	Pt^{4+}	6
V^{2+}	6	Sc^{3+}	6	Pd^{4+}	6
Fe^{2+}	6	Cr^{3+}	6	Nb^{5+}	7
Co^{2+}	4,6	Fe^{3+}	6	Mo^{5+}	8
Ni^{2+}	4,6	Co^{3+}	6	Hg^{2+}	4

The coordination number is previously considered to be a fixed number for a particular metal but many complexes are known in which the metal ion has more than one coordination number. Some examples are tabulated above :

The maximum coordination number of elements in the second row of elements of the periodic table is four, for the elements in the third and fourth rows it is six, and for the elements in the fifth or sixth row, six or eight are more commonly seen and in some cases it is ten. For the seventh row of the periodic table there seems to be some possibility of coordination number of twelve.

FACTORS AFFECTING THE COORDINATION NUMBER AND GEOMETRY OF THE COMPLEX

The coordination number of a metal ion depends on its nature, its oxidation state and on the ligands which are arranged around it. The coordination number is also influenced by the environmental factors such as temperature, pressure or solvent.

The geometry of the complex depends upon the coordination number of its central metal ion. If its coordination number is 6, the ligands are usually directed toward the corners of an octahedron and the shape of the complex is octahedral. Thus, it means that the ligands are coordinated to the central metal ion in a fixed geometry. Same is true for other coordination numbers.

Complex Ion

It is an electrically charged species which is formed by the union of a simple cation with one or more neutral molecules or one or more simple onions.

For example, $[Ni(NH_3)_6]^{2+}$ ion is obtained by the union of six molecules of ammonia with one Ni^{2+} ion. Similarly, $[Ag(CN_2)]^-$ is obtained by the union of two cyanide ions with one Ag^+ ion.

It is important to mention here that *the charge carried by a complex ion is equal to the algebraic sum of the charges carried by the central ion and the ligands attached to it.*

For example, the complex ferrocyanide ion, $[Fe(CN)_6]^{4-}$ has a charge of – 4 because the ferrous ion carries a charge of + 2 while six $(CN)^-$ ions carry a charge of – 6. In the case of complex ion $[Ag(CN_2)]^-$ has a charge of – 1 because Ag^+ ion has charge of + 1 and two cyanide ions have a charge of – 2. Complex ions are generally written inside the square brackets.

Coordination Sphere

The central metal atom and ligands which are directly attached to it are enclosed in square brackets and are collectively called as the coordination sphere.

The ligands and metal atom inside the square brackets behave as a single constituent unit.

ELECTRONEUTRALITY PRINCIPLES

The formation and stability of aquo and ammonia complexes can be explained on the basis of ion dipole electrostatic attraction but for ligands having negative character electroneutrality principle has to be invoked to explain their stability. Once the bond is formed there is no ditinction between a coordinate bond or a covalent bond for in both the cases the electrons are equally shared. Thus, in $[Co(NH_3)_6]^{3+}$ complex, out of 12 electrons shared between cobalt (III) and 6N atoms of ammonia molecule, the metal gains 6 electrons and so the formal charge changes from +3 to –3.

Thus, the formation of coordinate bonds implies the accumulation of negative charge on the metal atom. Earlier while discussing the Sidgwick's interpretation of Werner's compounds we have shown that coordinate bonds are formed by the donation of a pair of electrons by each ligand atom to metal ion. But Pauling has shown that such a situation need not arise which he expressed in his *electroneutrality principle.*

According to this principle *the electrons are distributed in a molecule in such a way as to make the residual charge on each of atom zero or very likely zero except that hydrogen and the most positive metals can acquire partial positive charge and the most electronegative atoms can acquire partial negative charge.* Thus obviously it implies that the M–L bond is not perfectly covalent. In other words, the bond is partially ionic and that the electron pair is attached more strongly to the ligand atom. Thus, if it is supposed that the M–L bond is 50 percent ionic, then the metal would be exactly neutral and the charge on each ligand atom would be +1/2. In addition Pauling has also suggested that accumulation of negative charge can be avoided if it is assumed that in addition to o bond, the central metal atom also forms n bonds by the back donation of electrons to the ligand atoms. Thus, the formation of

$$M \xrightarrow[\sigma]{\pi} L$$

π bond results in the decrease of negative charges on the metal atom. The *n* bond formation may take place in two ways :

(i) **dπ–pπ bonding :** *The first atom in a group which acts as donor atoms, (e.g., N,C in NO_2^- and CN^- respectively) do not have d-*

orbitals available for bonding. So the electrons donated by the filled dπ orbitals (dxy, or dyl orbitals) are accommodated in the vacant p orbitals of the ligand atom. Thus, there is an effective dπ–pπ overlapping between metal and ligand orbitals.

(ii) ***dπ-dπ bonding*** : *In this type of bonding the back donated electrons from dπ orbitals of the metal are accommodated in the dπ orbitals of the donor atoms. Such donor atoms, naturally will have vacant dπ orbitals and belong to the second short period or of the later periods e.g., P, S, As in* PR_3, SR_2 *and* AsR_3, *respectively.*

The formation of π bonding not only avoids the accumulation of negative charges on the metal atom but also strengthens the σ bond which explains the stability of many transition metals complexes with very poor electron pair donor atoms, such as CO, PX_2, PR_2, SR_2, and

THEORY OF CRYSTAL FIELD

This theory of Bethe and Van Vleck and applied mainly to ionic crystals, and is, therefore, called *Crystal Field Theory* (1956). Later *Orgel* was largely responsible for developing its general chemical aspects. The crystal field theory is mainly concerned with the interaction of d-orbitals of central metal with the surrounding ligands that produce crystal field effects. The crystal field theory is based on the following assumptions:

(i) The metal ion and ligands act as point charges and the interaction between them is purely electrostatic, *i.e.*, metal-ligand bonds are 100 per cent ionic.

(ii) The d–orbitals on the metal which were all of the same energy, *i.e.* degenerate in a free atom, have their degeneracy destroyed by the ligands when a complex is formed.

Salient Features of Crystal Field Theory

The various postulate are summarised below

(a) A complex is considered to be a combination of central metal ion surrounded by various ligands.

(b) The ligands in the complexes to be considered are either negatively charged ions *e.g.*, F and CN^- or neutral molecules, *e.g.* H_2O and NH_2, which have a dipole such that the negative ion is closest to the central ion.

(c) The attraction between central metal and ligand in a complex is regarded as purely electrostatic. The interaction between the electrons of the cation and those of the ligands is purely repulsive. It is these repulsive forces that are responsible for causing the splitting of the d-orbitals of the metal cation into two groups, t_{2g} and *eg.* The splitting depends on whether the ligands are arranged in an octahedral, tetrahedral or squareplanar way around the central ion. This effect is known as *crystal field splitting.*

(d) The electrons of the metal ion are repelled by the negative field of the ligands and therefore the metal electrons will occupy those d-orbitals which have their lobes farthest way from the direction of the ligand.

(e) In this theory, the covalent character of bond between metal and ligand is not taken into account.

(f) This theory does not permit the electrons to enter the metal orbitals. It means that it does not consider any orbital overlap.

(g) The number of ligands and their arrangement around the central ion will determine the crystal field.

Crystal field splitting of d-orbitals

In an isolated gaseous metal ion, all the five d orbital are degenerate, *i.e.*, have the same energy. The electrons which occupy such orbitals, do so according to the Hund's rule of maximum multiplicity.

Crystal field theory is basically concerned with the effect of different arrangement of surrounding ligands on the energy of the d-orbitals, and the subdivision of the orbitals into two groups, t_2g and e_g.

On the approach of the ligands, the electrons in the d-orbital of the central ion are repelled by the lone pairs of the ligands. This repulsion will raise the energy level of the d-orbitals.

All the ligands approaching the energy of each orbital will increase by the same amount. In other words they will remain degenerate, although they will have now higher energy than before.

But actually the d-orbitals differ in their orientation. Thus the energy of the orbitals lying in the direction of the ligands is raised to a larger extent than that of the orbitals lying in between the ligands. It means that the five degenerate d-orbitals of the metal ion will split up into two sets of orbitals having different energies.

This splitting of five degenerate d-orbitals of the metal ion under the influence of approaching ligands, into two sets of orbitals having different energies, is called *crystal field splitting or energy level splitting.*

The ligand field splitting depends on whether the ligands are arranged in an *octahedral, tetrahedral or square planar* way around the central ion.

The magnitude of the crystal field splitting of the d-orbitals energies may be represented by the symbol Δ. It is used to denote energy difference (the total energy difference may be regarded as made up of the energy lost or released in the lower levels together with the energy gained in the upper levels).

The crystal field splitting energy, Δ, is frequently measured in terms of a parameter, Dq. The magnitude of splitting is arbitrarily set at 10 Dq, *i.e.*, 2 Dq for each unpaired d–electron.

CRYSTAL FIELD SPLITTING BY DIFFERENT GEOMETRICALLY ARRANGEMENTS OF LIGANDS

The way in which the degeneracy of d-orbitals is resolved is different in octahedral, tetrahedral and square planar fields. We shall discuss separately the splitting in different fields.

Octahedral Complexes

In the octahedral complex, the six ligands are arranged octahedral around a central metal ion. In such an arrangement as the $dx^2–y^2$ and dz^2 orbitals lie along the *x, y* and *z* axes and so point directly towards the ligands, they experience much more repulsion than the remaining d-orbitals (*dxy*, *dyz* and *dzx*) which are directed in between the or, y and z axes.

Consequently, the energy of $dx^2 - y^2$ and dz^2 orbitals is increased much more in comparison to other d-orbitals. Thus, the d-subshell is splitted up into two degenerate sets, one consisting of more stable (lower energy) orbitals *dxy, dyz* and *dzx* and the other, less stable higher energy) orbitals $dx^2 - y^2$ and dz^2. These are commonly referred to as t_{2g} and e_g or d_ϵ and d_γ sets of orbitals respectively.

Degeneracy of all the five d–orbitals in the isolated central ion is represented by state I. The state II represents degeneracy (hypothetical) of all the five orbitals at a higher energy level. The state III represents *crystal field splitting*. The crystal field splitting is thus the energy difference between t_{2g} and e_g orbitals. It is denoted by the symbol Δ_0 or 10 Dq. (The subscript 0 indicated octahedral complexes). The geometry of an octahedral system shows that the energy of the t_{2g} orbitals is 2.5 Δ_0 or 0.4 Δ_0 or 4 Dq less than that of the hypothetical degenerated d orbitals which results in crystal field splitting is ignored. The energy of the e_g orbitals is 3/5 Δ_0 or 0.6 Δ_0 or 6 Dq above that of the hypothetical degenerate d–orbitals.

An electron prefers to move into an orbital of a lower energy. It is obvious that if an octahedral complex contains one d-electron, this electron will occupy one of the t_2g orbitals. The t_2g orbital has an energy 4 Dq less than that of the degenerate orbitals. Thus the complex will be 4 Dq more stable than predicted by the purely electrostatic theory which does not recognise any splitting of energy levels. This value, 4 Dq, is called the *crystal field stabilising energy* (C.F.E) of the complex under discussion.

If for each electron entering into a tag orbital, the crystal field stabilising energy is 4 Dq. then for each electron entering into an e_g orbital, the *destabilising energy is* 6 Dq. In other words, the stabilization energy assigned for an electron in e_g orbital is – 6 Dq. Working on this basis, it is possible to calculate crystal field stabilization energies for various metal ions in octahedral complexes.

Distribution of Electrons in *d* Orbitals and CFSE of Octahedral Complexes

The distribution of electrons in the *d* orbitals is based upon the following rules :

(i) *Electrons prefer to occupy the orbitals of lowest energy.*

(ii) *Electrons obey Hund's rule, i.e., electrons pairing will not take place until all the available orbitals of a given set contain one electron each.*

(iii) *If the energy to place an electron in the upper e_g level is greater than the energy to pair electrons (pairing energy = P) in the lower t_2g level* (10 Dq < P), *then the electrons will pair up in t_2g orbitals.*

(iv) *If the energy to place an electron in the upper eg level is less than the pairing energy in the lower t_2g orbital* (10 Dq < P), *then the electrons will occupy e_g orbitals.*

Weak and Strong Field Cases

Ligands which cause only a small degree of splitting of *d* orbitals, *i.e.,* weak field, are called *weak field ligands* and those which cause a large splitting. *i.e.,* strong field, are called *strong field ligands.* The magnitude of 10 *Dq* and hence the distribution of the electron in *d* orbitals are very much influenced by the strength of the ligand field. The CF splitting ability of the ligands decreases in the following order :

$$CO > CN^- > NO_2^- > en > NH_3 > H_2O > F^- > OH^- > Cl^- > Br^- > I^-$$

Strong field ligands　　　　　　　　　Weak field ligands

In a weak field, 10 Dq is generally less than pairing energy, *i.e.,* 10 Dq < P. It means that electrons obey Hund's rule and remain unpaired. But in the case of a strong field, 10 Dq > P; it follows that electrons pair up in the lower energy orbitals. We shall now consider the distribution of *d* electrons in octahedral complexes, in the presence of weak field and strong field ligands.

In octahedral complexes with d^1 configuration, the *d* electron will occupy the lower t_{2g} orbitals irrespective of the field strength and the crystal field stabilisation energy is – 4Dq. For d^2 configuration, the two electrons will occupy the lower two t_{2g} orbitals and CFSE is – 4Dq × 2 = 8 Dq. Similarly for d^3 configuration CFSE is – 12Dq and the t_{2g} level is just half-filled. Thus complexes with d^1, d^2 and d^3 configuration, irrespective of the ligand field strength, remain paramagnetic, with one, two and three unpaired electrons respectively.

For d^4 configuration, two possibilities will arise, that is, the fourth electron can either enter the upper e_g level or pair up in the lower t_{2g} level. Which of these two arrangements occurs depends upon the values of 10 Dq and P, which in turn depends upon the strength of the ligands. In a *weak field,* 10 Dq < P, hence the fourth electron prefers to enter one of the e_g orbitals rather than pairing in t_{2g} orbitals. This way, it loses somewhat less energy in the form of CFSE by entering the destablised e_g level. The net CFSE energy is :

$$CFSE = (3 \times -4\ Dq) + 1 \times (+6Dq) = -6Dq$$

In a *strong field* where 10 Dq > P, the fourth electron prefers to pair up in one of the three lower t_{2g} orbitals rather than entering the upper e_g orbitals. CFSE for this type of complex is (4 × – 4Dq + P) because all the four electrons enter t2g orbitals and some energy is used to pair up the electron.

Thus, for d^4 configuration the electronic arrangement' differs in weak and strong fields. In weak field there are four unpaired electrons while in the strong ligand field, there are only two unpaired electrons. These two arrangements thus differ in the number of unpaired electrons. The arrangement with maximum unpaired electrons is called *'high spin'* or *'spin free'* and the one with minimum number of unpaired electrons is called *low spin or spin paired* arrangement.

For d^5 configuration, in a weak field three electrons occupy the t_{2g} orbitals and the other two in two e_g orbitals (one in each). CFSE is zero because the presence of two electrons in the unfavourable e_g level exactly balances, the stabilisation from the three electrons in the t_2g level. In other words, the half filled shell (d^5) is physically symmetrically and no stabilisation can occur through the application of external ligand field. In case of *strong field* the fifth electron will pair up in the t_{2g} orbital.

$$\therefore \qquad \text{CFSE} = -20\ \text{Dq} + 2\text{P}$$

For d^6 configuration in a weak field, the sixth electron will pair up in the lower t_{2g} orbitals because all the five d-orbitals are already half-filled CFSE = – 4Dq + P. In case of *strong field,* the sixth electron will pair up in the t_{2g} orbitals. Thus, all the three t_2g orbitals will have paired electrons.

$$\therefore \qquad \text{CFSE} = -24\ \text{Dq} + 3\text{P}$$

Similarly, *for* d^7, d^8 and d^9 and d^{10} configurations, the d–electrons of the metal can be distributed and their CFSE can be calculated.

It is clear that for d^1, d^2,d^3, and d^8, d^9, d^{10} configurations, the distribution of electrons and CFSE are the same irrespective of the strength of the ligand field. But for configuıations d^4 to d^7, two arrangements (high-spin and low- spin) are possible depending upon the strength of the ligand field.

For maintaining the average constant, t_{2g} orbitals are raised to 4 Dq while e_g orbitals are lowered by 6 Dq from the barycenter. Thus, *the energy level scheme for the tetrahedral symmetry is exactly the reverse of that for octahedral symmetry.*

It has been observed that the energy separation between the e_g and t_2g levels Δ_1, in tetrahedral symmetry is smaller than in octahedral symmetry Δ_0 (10 Dq). This observation is as expected because the d-orbitals are not directly affected by the ligands and moreover the number of ligands is smaller. For equivalent ligand charges and distances :

$$\Delta_t = -\frac{4}{9} D^\circ$$

where the minus sign means the reversal of the orbital levels. Because of the low value of Dq in tetrahedral symmetry *crystal field favours the formation of octahedral complexes than tetrahedral complexes.*

Low spin (strong field) tetrahedral complexes have not yet been obtained, probably because of crystal field stabilisation energies are never sufficient to bring out spin pairing.

Thus. *only high spin* or *weak field* cases are to be considered for tetrahedral complexes. This simplifies the treatment of electron configuration and CFSE.

In a weak field, pairing energy (P) is more than Δ_t. Hence, electrons enter even the higher levels and t_{2g} remain unpaired until the sixth electron forces pairing. For example in a complex with d^4 configuration two electrons occupy the lower e_g orbitals and two occupy the higher t_{2g} orbitals, all the four electrons remaining unpaired.

Hence, CFSE = – 4Dq *i.e.*,

$$(2 \times -6Dq) + (2 \times 4Dq) = -4Dq$$

The distribution of electrons and the crystal field stabilisation energies for d^1 to d^{10} configuration are given in Table 3.

Table 3 : Distribution of D electrons for weak field in tetrahedral complexes

d^n	*Distribution of electrons* e_g	t_{2g}	*No. of unpaired electrons*	*CFSE*
d^1	↑		1	– 6Dq
d^2	↑ ↑		2	– 12Dq
d^3	↑ ↑	↑	3	– 8Dq
d^4	↑ ↑	↑ ↑	4	– 4Dq
d^5	↑ ↑	↑ ↑ ↑	5	– 0Dq
d^6	↑↓ ↑	↑ ↑ ↑	4	– 6Dq + P
d^7	↑↓ ↑↓	↑↓ ↑ ↑	3	– 12Dq+2P
d^8	↑↓ ↑↓	↑↓ ↑ ↑	2	– 8Dq + 3P
d^9	↑↓ ↑↓	↑↓ ↑↓ ↑	1	– 4Dq + 4P
d^{10}	↑↓ ↑↓	↑↓ ↑ ↑ ↑↓	0	– 0Dq + 5P

Square Planar Complexes

In the square planar complexes due to the absence of the ligand groups along the z-axis, the dz^2 orbitals becomes most stable and $dx^2 - y^2$ least stable.

However, most of the square planar complexes are such, where the two ligands along the z-axis are not totally removed but placed at a longer distance than the ligands along the x and y axes (*e.g.*, Cu^{2+} and Ni^{2+} complexes). In such cases the *dz* orbital though becomes more stable than $dx^2 - y^2$ and d_{xy} orbitals, it remains less stable in comparison to *dxz* and *dyz* orbitals, which still form a degenerate set. The sequence of orbitals in the increasing order of energy would then be *dxz* and *dyz*, dz^2, *dxy* and $dx^2 - y^2$.

FACTORS AFFECTING Δ

The magnitude of Δ or 10Dq is influenced by a number of factors such as :

Position of the Metal in Transition Series

This in an important factor and the magnitude of crystal field splitting depends upon the position of metal in transition series, *i.e.*, whether it belongs to first, second or third series. In general, the value of Δ_0 or 10 Dq increases on descending a group of transition element, *i.e.*, from first to third transition series. This point has been illustrated in Table 4.

Table 4

Complex	Δ_0 cm^{-1}	Δ_0 *(Kcal/mol)*
$[Co(NH_3)_6]^{3+}$	23,000	65.70
$[Rh(NH_3)_6]^{3+}$	34,000	97.14
$[Ir(NH_3)_6]^{3+}$	41,000	117.1

Nature of the Ligands

If we compare the absorption spectra of various transition metal complexes, we observe that the position of absorption spectra and hence the value of Δ or 10 Dq for any given metal ion depends upon the nature of the ligands attached. In order to illustrate this point, we have given two values of Δ_0 or 10Dq for a number of Cr^{3+} complexes with different ligands in Table 5.

Table 5

Complex	*Absorption peak* cm^{-1}	*10 Dq or* Δ_0 *(Kcal/mol)*
$[CrCl_6]^{3-}$	13640	38.96
$[Cr(H_2O)6]^{3+}$	17830	50.94
$[Cr(NH_3)_6]^{3+}$	21680	61.94
$[Cr(CN)_6]^{3-}$	262.0	74.08

From the above table we can draw the important conclusion that the cyanide ligand produces more splitting and is therefore a strong ligand whereas chloride ligand produces less splitting and is therefore a weak ligand. Similarly, we can compare the various ligands, and then arrange them in the order of their strength, in the form of a series called spectrochemical series.

$$I^- < Br^- < S^{2-} < SCN^- < Cl^- < NO^{3-} < F < OH^-$$
$$< H_2O < NCS^-, NH_3 < en < dipy < NO_2^- < CN^- < CO.$$

Charge on the Metal Ion

As earlier stated that crystal field theory is based on electrostatic model. It means that the ionic charge present on the central metal has a direct effect on the magnitude of 10Dq.

A generalisation is that a metal with higher charge produces more splitting than an ion with lower charge. For example, Δ_0 or 10 Dq for hexa-aquo complexes of Cr^{2+} and Cr^{3+} are 39.7 Kcal mol^{-1} and 51.94 Real mol^{-1}.

Geometry of the Complex

In octahedral complexes, the splitting of *d* orbital has been found to be more than twice than in tetrahedral complexes, for the same metal ion and ligands. This difference in Δ value is due to the following facts.

(i) *In octahedral complexes six ligands are involved whereas in tetrahedral complexes only four ligands are involved. This results in 33 per cent decrease in the field strength, provided the other factors do not change.*

(ii) *In octahedral complexes ligands are positioned exactly in the direction of* d_z^2 *and* $dx^2 - y^2$ *orbitals whereas in tetrahedral complexes, the ligands are not positioned at any of the d-*

orbitals and thus have more influence on the t_{2g} *orbitals than on the* e_g *orbitals. However, in case of square planar complexes, the degree of splitting is more than in a tetrahedral field.*

MEASUREMENT OF Δ OR 10 Δq

The splitting of d-orbitals by a ligand field into two sets with different energies provides a possible electronic energy change within a complex. An electron in an orbital of lower energy might be excited into an orbital of higher energy by the absorption of radiation. It so happens that the radiation necessary for such excitation lies in the visible region of light and this explains why many complexes are coloured. Study of absorption spectra of complexes enables Δ values to be obtained. A band or bands, in the absorption spectra, can be related to the excitation of an electron or electrons, between d-orbitals splitted by a ligand field. The ability to interpret such spectroscopic evidence is one of the advantages of crystal field theory. The situation in the $[Ti\ (H_2O)_6]^{3+}$ ion, hexaquotitanium (III) ion, is particularly simple for thcrc is only one *d* electron (d^1) involved. In a ground state, it will occupy one of the t_{2g} (d_ϵ) orbitals but absorption of radiation will excite it into one the e_4 ($d\lambda$) orbitals.

The absorption spectrum of $[Ti\ (H_2O_6]^{3+}$ shows a strong band corresponding to wavelength of 5000A°. This represents an energy change of 57.15 Kcal/mole which must be the Δ value for this complex. Light of wavelength 5000A° is green. If such a light is absorbed, the transmitted light appears blue and red, *i.e.*, purple, and $[Ti(H_2O)_6]^{+3}$ ion is in fact, purple. Table 1.9 gives a few of the values of Δ for the various transition metal ions and ligands. While value of Δ is approximately constant for ions of a given charge with the same ligand, changing the ligand does change A and thus alters the absorption spectrum associated with the metal ion.

2

Application of Coordination Compounds

INTRODUCTION

Profound changes occur in the properties of a metal ion on complex formation. These changes are generally reflected in solubility, colour, stability, magnetic properties, etc. These changes in properties find manifold applications.

As catalysts : Many enzymes, which serve as the catalysts in living system, are coordination compounds. For example, the decomposition of hydrogen peroxide is catalysed by many things, including iron compounds.

$$2H_2O_2 \xrightarrow{\text{catalyst}} 2H_2O + O_2.$$

This is evident from the following points

(i) *Ordinary hydrated ferric ion has a relative activity of unity,*

(ii) *A co-ordination compound of iron, the he me (human blood), has a relative activity of one thousand, and*

(iii) *Catalase, a heme surrounded by a complicated protein structure, has a relative catalytic activity of ten billion.*

Dyes : The characteristic colours of co-ordination compounds would indicate their uses as dyes. Alfred Werner demonstrated the relation of co-ordination with dyeing and he showed that several compounds which were capable of forming metal-chelate compounds were able to dye cloth pretreated with ferric hydroxide.

The first complete study of co-ordination compounds as dyes was given by G.T. Morgan and his co-workers in the early 1920.

Analytical Applications : The basis of qualitative detection of metal ions is the complexation reactions of such ions.

Common examples of qualitative detection of metal ions by way of complexation reactions include detection of nickel (II), as red coloured bis (dimethylglyoximato) nickel (II), of cobalt as blue coloured tetrakis (isothiocyanato) cobalt (II), $[Co(NCS)_4]^{2-}$, of iron as blood red hexakis (isothiocyan-to) iron (III), $[Fe(NCS)]_3^-$ etc.

These are as follows.

(a) Aluminium and zinc salt solutions form with excess of NaOH, soluble sodium aluminate and sodium zincate.

$$AlCl_3 + 3NaOH \longrightarrow \underset{\text{ppt.}}{Al(OH)_3} + 3NaCl$$

$$Al(OH)_3 + NaOH \longrightarrow \underset{\text{Sod. alu min ate}}{NaAlO_2} + 2H_2O$$

$$ZnCl_2 + 2NaOH \longrightarrow \underset{\text{ppt.}}{ZnOH_2} + 2NaCl$$

$$Zn(OH)_2 + 2NaOH \longrightarrow \underset{\text{Sod. zincate}}{Na_2\,ZnO_2} + 2H_2O$$

Similarly, chromium hydroxide dissolves in excess of NaOH forming sodium chromite.

$$CrCl_3 + 3NaOH \longrightarrow Cr(OH)_3 + 3NaCl$$

$$2Cr(OH)_3 + 2NaOH \longrightarrow \underset{\text{Sod. chromite}}{2Na_2Cr_2O_4} + 4H_2O$$

Chromium salts in the presence of oxidising agents such as Br_2 water, H_2O_2, form a soluble yellow chromate ion.

$$2CrCl_3 + 10NaOH + O \longrightarrow \underset{\text{Sod. chromate}}{2Na_2CrO_4} + 6NaCl + 5H2O$$

These reactions have been utilised for the separation of aluminium from iron and chromium and zinc from manganese.

(b) Iodine is having feeble solubility in water. However, its solution is prepared by dissolving iodine in a solution of potassium iodide. This solubility is due to the formation of complex KI_3.

$$KI + I_2 \longrightarrow KI_3 \rightleftharpoons K^+ + I_3^-$$

When potassium iodide solution is added to mercuric chloride solution, a scarlet red ppt. of mercuric iodide is obtained which dissolves in excess of KI forming K_2HgI_4.

$$HgCl_2 + 2KI \longrightarrow HgI_2 + 2KCl$$

$$HgI_2 + 2KI \longrightarrow \underset{\text{(Complex)}}{K_2HgI_4}$$

When this complex is added to sodium hydroxide, the solution is known as Nessler's reagent which is employed for the *detection of ammonia.* It forms brown colour or precipitate with ammonia.

$$2K_2HgI_4 + 3NaOH + NH_4OH \rightarrow \begin{matrix} NH_3 \\ \quad Hg \\ O \\ \quad Hg \\ \\ I \end{matrix} + 4KI + 3NaI + 3H_2O$$

Iodine of Millon's base

(c) When potassium cyanide is added to copper and cadmium salts solution, both copper and cadmium from complexes. These two complexes have different stabilities.

$$CuSO_4 + 2KCN \longrightarrow \underset{\text{ppt.}}{Cu(CN)_2} + K_2SO_4$$

$$2Cu(CN)_2 + 6KCN \longrightarrow \underset{\text{Pot. curprocyanide}}{2K_3[Cu(CN)_4]} + (CN)_2$$

$$CdSO_4 + 3KCN \longrightarrow Cd(CN)_2 + K_2SO_4$$

$$Cd(CN)_2 + 2KCN \longrightarrow \underset{\text{Pot. cadmicyanide}}{K_2[Cd(CN)_4]}$$

When hydrogen sulphide is passed in these two solutions of complexes, cadmium is only precipitated because the copper complex is much more stable than the cadmium complex.

$$K_3Cu(CN)_4 \rightleftharpoons 3K+ + [Cu(CN)_4]^{3-}$$

$$[Cu^+ + (CN)_4]^{3-} \rightleftharpoons Cu^+ + 4CN^- \text{ (Negligible dissociation)}$$

$$K_2[Cd(CN)_4] \rightleftharpoons 2K^+ + [Cd(CN)_4]^{2-}$$

$$[Cd(CN)_4]^{2-} \rightleftharpoons Cd^{2+} + 4CN^- \text{ (High dissociation)}$$

(d) **Separation of nickel and cobalt :** When potassium cyanide is added to a mixture of cobalt and nickel salts, potassium cobalto cyanide is formed whereas nickel only forms nickel cyanide. Potassium cobalto cyanide with bromine and alkali is converted into potassium cobalticyanide whereas nickel cyanide with bromine and NaOH yields a black ppt. of nickel oxide.

$$CoCl_2 + 2KCN \longrightarrow Co(CN)_2 + 2KCl$$

$$Co(CN)_2 + 4KCN \longrightarrow K_4[Co(CN)_6]$$

$$2K_4[Co(CN)_6] + H_2O + O \longrightarrow 2K_3[Co(CN)_6] + 2KOH$$

$$NiCl_2 + 2KCN \longrightarrow Ni(CN)_2 + 2KCl$$

(e) **Reaction with dimethyl glyoxime :** Ni^{2+} form a red ppt. with dimethyl glyoxime. This reaction has been used for the separation of nickel from cobalt in the group IV.

$$2\begin{array}{l} CH_3 - C = N - OH \\ \quad\quad | \\ CH_3 - C = N - OH \end{array} + 2NH_4OH + NiCl_2 \longrightarrow$$

Dimethy lg lyoxime

(f) **Reactions with yellow ammonium sulphide :** The reaction of yellow ammonium sulphide with sulphides of IInd group cations is used for the separation of IIA from IIB. IIB cations form soluble complexes while of IIA cations do not form.

Arsenic,

$$As_2S_3 + 3(NH_4)_2S \longrightarrow \underset{\text{soluble}}{2(NH_4)_3AsS_3}$$

$$As_2S_5 + 3(NH_4)_2S \longrightarrow \underset{\text{soluble}}{2(NH_4)_3AsS_4}$$

Antimony,

$$Sb_2S_3 + (3NH_4)_2S \longrightarrow \underset{\text{soluble}}{2(NH_4)_3SsS_3}$$

$$Sb_2S_5 + 3(NH_4)_2S \longrightarrow \underset{\text{soluble}}{2(NH_4)_3SbS_4}$$

Tin

$$SnS + (NH_4)_2S \longrightarrow \underset{\text{soluble}}{(NH_4)_2SnS_2}$$

$$SnS_2 + (NH_4)_2S \longrightarrow \underset{\text{soluble}}{(NH_4)_2SnS_2}$$

These thioarsenite, thioarsenate, thioantimonite, thioantimonite, thiostannanite and thiostannate are soluble complexes from which original sulphides of arsenic, antimony and tin can be ppted by making the solution acidic with dilute hydrochloric acid.

(f) Complexes are usually formed to keep a metal ion in solution or to dissolve precipitate and bring the metal ion back into solution. For example, AgCl dissolves on addition of ammonia solution and forms the complex ion, diamine silver (I) ion.

$$AgCl + 2NH_3 \longrightarrow [Ag(NH_3)_2]^+ + Cl^-$$

(g) Formation of complexes with characteristically coloured solutions or precipitates is used for the detection of metal ion in solution. For example, Cu^{2+} ion can be detected in solution by the dark blue coloured solution it forms on addition of ammonia.

$$Cu^{2+} + 4NH_3 \longrightarrow \underset{\text{dark blue}}{[Cu(NH_3)_4]^{2+}}$$

(h) **Use in gravimetric determination :** Inner complexes are often insoluble in aqueous medium but soluble in organic solvents. The formation of such chelates often needs suitable pH range and many metal ions can be quantitatively precipitated and metal ions determined gravimetrically. Some applications include estimation of aluminium as yellow coloured tris (8–hydroxy quinolonate), aluminium (III) and copper (II) as light green coloured bis-(quinolonate) aluminium (III) and copper (II). Some complexes are unstable at the drying temperature (120–150°C) and such complexes are ignited to metal oxides.

(i) **Use of organic sequestering agents in removal of interference in gravimetric estimations** An organic ligand may read with more than one metal ion to form sparingly soluble precipitates. In such cases direct estimation of a particular metal ion in the presence of interfering ions is not possible. A strong sequestering agent like EDTA is of great importance in making interfering ions ineffective, and thus making the precipitating ligand almost specific for a particular metal ion. For example, in the presence of EDTA, beryllium may be precipitated with ammonia in presence of chromium, cobalt, cadmium, iron, copper, lead, manganese, zinc, aluminium, bismuth etc.

(j) **Solvent extraction and separation of metal ions through complexation :** Both iron (III) and aluminium (III) form complexes with 8-hydroxy quinoline. Iron (III) complex is easily soluble in chloroform. Iron (III) can be separated from aluminium (III) around pH 3.0 by the addition of 8–hydroxy–quinoline (oxine), followed by extraction with chloroform.

(k) **Complexometric titrations :** The sequestering effect of EDTA has led to the development of some versatile titrimetric reagents for the estimation of metal ions. Mostly EDTA titrations are performed with indicators, which are chelating agents whose metal complexes have different colours from the reagent. The indicators most commonly employed are Erichrome black T, murexide, etc. Such complexometric titrations are used in the determination of water hardness. This procedure affords various advantages as compared to old and laborious palmitate titrations.

(l) **Spectrophotometric estimation of micro amounts of metal ions.** A large number of metal complexes are intensely coloured which provide a basis for their spectrophotometric determination when present in trace amounts. For example, iron (II) forms dark red complex with o-phenanthroline over a wide range of pH 2–9. Trace amounts ($\sim 10^{-6}$ gm/ml.) of iron (II) can be estimated by measuring the optical density of the complex at 5.5 mμ.

Photography : Formation of $[Ag(S_2O_3)_2]^{3-}$ complex ion has an important role in photography. After the exposed film is developed, sodium thiosulphate solution is used to dissolve the unreduced silver halide.

$$AgBr + 2S_2O_3^{2-} \longrightarrow [Ag(S_2O_3)_2]^{3-} + Br^-$$

Extraction of Metals from Ores : The noble metals like Ag and Au are extratced from their ores through the formation of cyanide complexes like $[Ag(CN)_2]^-$ and $[AuCl_4]^-$ respectively.

Artificial Silk : Schweitzer's reagent, *i.e.,* tetrammine copper (II) hydroxide $[Cu(NH_3)_4](OH)_2$ is used as a solvent for cellulose during the manufacturing of artificial silk.

Biological Applications : A number of metal complexes are of biological importance. In particular (i) haemoglobin in the red blood cells contains an iron-porphyrin complex, (ii) chlorophyll in green plants contains magnesium-porphyrin complex, (iii) vitamin B^{12} is a cobalt complex. Body contains a number of compounds like adrenaline, citric acid and cortisone which form unwanted complexes with metals like lead, copper etc. and prevent normal metabolism. Lead poisoning and copper poisoning are treated by injecting EDTA, so that metal–EDTA complex is excreted in the urine.

CRYSTAL FIELD VALENCE AND BOND THEORY COMPARISON

The properties of complexes which could not be explained on the basis of valence bond theory are satisfactorily explained by crystal field theory. CFT is thus definitely an improvement over VBT. The following merits of CFT over VBT will prove that statement :

(i) CFT predicts a gradual change in magnetic properties of complexes rather than the abrupt change predicted by VBT.

(ii) In some complexes, when Δ_0 is very close to P, simple temperature changes may affect the magnetic properties of complexes. Thus, CFT provides theoretical basis for understanding and predicting the variation of magnetic moments with temperature as well as the detailed magnetic properties of complexes. This is just in contrast to VBT which cannot predict or explain magnetic behaviour beyond the level of specifying the number of unpaired electrons.

(iii) Though the assumptions inherent in VBT and CFT are vastly different, the main difference lies in their description of the orbitals not occupied in the low spin states. VBT forbids their use as they are involved in forming hybrid orbitals, while CFT strongly discourages their use as they are repelled by the ligands.

(iv) According to VBT, the bond between the metal and the ligand is covalent, while according to CFT it is purely ionic. The bond is now considered to have both ionic and covalent character. Unlike valence bond theory.

(v) CFT provides a framework for the ready interpretation of such phenomena as tetragonal distortions.

(vi) CFT provides satisfactory explanation for the colour of transition metal complexes, *i.e.*, spectral properties of complexes.

(vii) CFT can semiquantitatively explain certain thermodynamic and kinetic properties.

(viii) CFT makes possible a clear understanding of stereochemical properties of complexes.

Ligand Field Theory

Recently based on crystal field theory of *Bethe* (1929) and *Van Vleck* (1931–1935), a new approach known as ligand field theory was postulated.

In the crystal field theory, attraction between central metal and ligand in a complex is regarded as purely elctrostatic, *i.e.,* the bonding between the central metal ion and ligands is purely ionic. Thus, in crystal field theory covalent character of bond between metal and ligand is not taken into account. However, the enough evidence is available which suggests there is some measure of covalent bonding in complexes. The various experimental evidences are as follows.

(i) *Most direct evidence is obtained from electron* ***resonance spectrum*** *(ESR) of complexes. ESR spectrum of* $[IrCl_6]^{2-}$ *clearly shows hyperfine splitting indicating the delocalization of d electrons onto six chlorines. Such study of other complexes too gives similar results.*

(ii) ***Nuclear magnetic resonance (NMR)*** *studies of the complexes like* $KMnF_3$ *and* $KNiF_3$, *definitely indicate that the metal* t_{2g} *and* e_g *electrons pass a fraction of time around the fluorine nuclei.*

(iii) ***Nuclear quadrupole resonance (NQR)*** *spectra of square planar halide complexes of divalent platinum and palladium suggest considerable covalency in the metal-ligand bonds.*

(iv) ***The unusually large absorption band intensities*** *observed for tetrahedral complexes like* $[CoCl_4]^{2-}$ *have been interpreted as due to appreciable covalent nature of the metal-ligand bonds.*

The ligand field theory is eventually the same as pure crystal field theory but covalent character being taken into account.

When the orbitals overlap, *i.e.,* covalent character is excessive as in metal complexes of carbon monoxide or the isocyanide, the *molecular orbital theory* gives a more complete explanation of the *metal-ligand bonding.*

Ligand field theory differs from the valence bond theory by recognising the existence of antibonding as well as the bonding and non-bonding orbitals. For the present, the ligand field theory seems to offer the most practical approach to bonding in co-ordination compounds.

In ligand field theory, it is assumed that molecular orbitals are formed by the overlap of orbitals from the ligand with the atomic orbitals of the central atom. (The procedure is similar as adopted in case of simple molecules). The molecular orbitals thus formed may be of a *bonding, antibonding or a non-bonding character.* The antibonding orbitals are similar to the bonding orbitals except that the antibonding orbitals lie higher in energy and have nodes or regions of low electron density between the central atom and the ligands. The antibonding orbitals are of interest because it is these into which electrons may be excited from the *tis* orbitals by absorptions of energy. The non-bonding orbitals in octahedral complexes are simply the *dxy, dxz* and *dyz* orbitals of the central atom.

As the number of molecular orbitals formed is always equal to the number of atomic orbitals taking part in the overlappings and this number is quite large for complexation processes. The MO energy level diagrams for metal complexes are highly complicated. The fact that σ as well as π (both types of overlaps are involved in the coordination process) add more complication to these diagrams.

The three common types of complexes are octahedral, tetrahedral and square planar. But we shall discuss octahedral complexes.

CONFORMATION ISOMERISM

In this isomerism two isomers have different geometries but are otherwise identical. For example $[Ni^{2+}(P.Et.Ph_2)_2Br_2]$ complex (four coordinated complex) gives two conformation isomers. One of these is green and paramagnetic while the other is brown and diamagnetic. The green form is tetrahedral while the brown form is square planar. Thus these have different geometries but the coordination number of Ni^{2+} ion in both the isomers is the same (=4). In other words in *compounds which have the same stoichiometric composition but yield different ions in solution are called ionisation isomers and the phenomenon is called ionisation isomerism.*

Such isomers are produced by an interchange of position of co-ordination ions (ligands) inside and outside the complex. Ionisation isomers are easily detected by the quick determination of conductance in solution and by chemical tests. For example, the violet bromopentaammine sulphate of cobalt with a formula of $[Co(NH_3)_4Br]SO_4$ is an ionisation isomer of the red sulphatopentammine bromide with a formula of $[Co(NH_3)_4SO_4]$ Br. Both of these compounds have the similar empirical formula, but the first complex reacts with $BaCl_2$ to precipitate immediately $BaSO_4$ and no precipitate with $AgNO_3$. On the other hand, sulphatopentammine cobalt gives a precipitate of AgBr with $AgNO_3$ and no precipitate with $BaCl_2$. Other illustrations of ionisation isomerism are

$[Co(NH_3)_4Cl_2]NO_2$ and $[Co(NH_3)_4Cl(NO_2)]Cl$

$[Pt(NH_3)_4(OH)_2]SO_4$ and $[Pt(NH_3)_4SO_4](OH)_2$

$[Pt(NH_3)_4Cl_2]Br_2$ and $[Pt(NH_3)_4Br_2]Cl_2$.

IONISATION ISOMERISM

The complex compounds showing ionisation isomerism are called *ionisation isomers*. These isomers have the same molecular formulae and are produced when the ligands given in the coordination sphere and the anions present out side the coordination sphere are exchanged with each other. Thus both the complex compounds of each of the following pairs are ionisation isomers to each other.

(i) $[Co(NH_3)_5Br]SO_4$ and $[Co(NH_3)_5(SO_4)]Br$

(ii) $[Co(NH_3)_4Cl_2]NO_2$ and $[Co(NH_3)_4Cl(NO_2)]Cl$

(iii) $[Pt(NH_3)_4Cl_2]Br_2$ and $[Pt(NH_3)_4Br_2]Cl_2$

(iv) $[Co(NH_3)_5(NO_3)]SO_4$ and $[Co(NH_3)_5(SO_4)]NO_3$

(v) $[Cr(NH_3)_4I_2]Br$ and $[Cr)NH_3IBr]I$

(vi) $[Fe(NH_3)_5(CN)]SO_4$ and $[Fe(NH_3)_5SO_4]CN$.

Ionisation isomers show different properties, *e.g.*, $[Co(NH_3)_5Br]SO_4$ (I) and $[Co(NH_3)_5(SO_4)]Br$ (II) which are ionisation isomers to each other show different properties as mentioned below:

(i) (I) is red-violet while (II) is red.

(ii) When these isomers are dissolved in water, they give different ions. For example:

$$[Co(NH_3)_5Br]SO_4 \rightleftharpoons [Co(NH_3)_5Br]^{2+} + SO_4^{2-}$$
(I) (red-violet)

$$[Co(NH_3)_5SO_4]Br \rightleftharpoons [Co(NH_3)_5SO_4]^{+} + Br^{-}$$
(II) (red)

The compounds which possess the same chemical composition but have different structural arrangements of their atoms are called isomers and the phenomenon is known as isomerism.

The phenomenon of isomerism is fairly common in organic compounds but it is less common in inorganic compounds. However, coordination compounds provide conditions which not only favour the usual isomerism but also give rise to unusual isomerism which is only found in coordination compounds Such an isomerism arises due to :

(*i*) *variety of bonds,*

(*ii*) *multiplicity of molecular arrangements, and*

(*iii*) *complexity of stereochemical relationships.*

The various types of isomerism are outlined below :

Hydrate Isomerism : *This type of isomerism arises due to different positions of a water molecule in a complex. For example,* three isomers with the formula $CrCl_3$. $6H_2O$ are known.

(i) $[Cr(H_2O)_6]Cl_3$ Violet Hexa–aquochromium (III) chloride

(ii) $[Cr(H_2O)_5Cl]Cl_2[H_2O]$ Light green Chloropentaquochromium (III) chloride

(iii) $[Cr(H_2O)_4Cl_2]Cl[_2H_2O]$ Green Dichlorotetraaquochrormium (III) chloride

The violet coloured compound : (i) does not lose any water molecule on dehydration over sulphuric acid and all the three Cl^- ions are precipitated by silver nitrate as silver chloride. The compound (ii), light green, loses one water molecule on dehydration over sulphuric acid and only two Cl^- ion are precipitated as silver chloride. The compound (iii), dark green, loses two water molecules on dehydration over sulphuric acid and only one chloride ion is precipitated as silver chloride. Another example of hydrate isomerism is

$$[Co\ (NH_3)_4\ (H_2O)\ Cl]\ Cl_2\ ;\ [Co(NH_3)4Cl_2]Cl.H_2O.$$

Co–ordination Position Isomerism : *This type of isomerism occurs in polynuclear complexes in which co-ordinating groups may be present in the same number but may arrange themselves differently with respect to the different metal ions present, i.e.,* it is a result of the exchange of ligands between the different metal nuclei. An example of this is

Thus, ammonia molecules and chloride ions are differently placed relative to the two cobalt ions.

POLYMERISATION ISOMERISM

This term is used to describe compounds which have the same stoichiometric composition but whose molecular compositions are

multiplies of the simplest stoichiometric arrangement. An example is

$$[Pt(NH_3)_2Cl_2].Pt(NH_3)_4][PtCl_4],[Pt(NH_3)_3Cl]_2[PtCl_4]$$

I II III

In the above example, the second and third compounds may be said to polymers of the first. Actually, in the strict sense of the word, these are not examples of polymerisation. Polymerisation is usually interpreted as meaning a multiplication of simple molecules. Here we have differences in arrangements as well as a multiplication of molecular weight.

Another example illustrating polymerisation isomerism is:

$$[Cr(NH_3)_3(CNS)_3] \text{ and } [Cr(NH_3)_5(CNS)]_3\ [Cr(CNS)_6]_2$$

6. Structural Salt or Linkage Isomerism: *This type of isomerism occurs when more than a single atom in a coordinating group may function as a donor.* A number of groups might cause isomerism of this type. Thus,

(a) The thiocyanate radical with the structure SCN^- might co-ordinatc either through the sulphur or the nitrogen. The isomers having this type of isomerism were prepared in 1963.

$$[Pd(dipy)\ (SCN)_2] \text{ and } [Pd(dipy)\ (NCS)_2]$$

(b) In the case of NO_2^- either nitrogen or oxygen atom can act as donor. Therefore (NO_2^-) ion may co-ordinate to the central ion either through oxygen or nitrogen giving two isomers called linkage or structural isomers :

The nitro group may link *via* nitrogen (nitro–N isomer) or *via* oxygen (nitro–O,) *e.g.*, yellow brown nitropentammine cobalt (III) chloride $[Co(NH_3)_5(NO_2)\ Cl_2$ and red nitrito pentammine cobalt (III) Chloride $[Co(NH_3)_5ONO)]Cl_2$. The red compound, either in solution or in the solid state, undergoes slow transformation into the yellowish-brown one. These two compounds were prepared by Jorgenson.

(c) Besides SCN^-, $S_2O_3^{2-}$ and NO_2^- groups, there are other ligands which are capable of playing double role, such as CN^-. $(NH_2)_2CO,(CH_3)_2SO$.

Linkage isomerism is considered to be a special case of ambidentate behaviour of ligands. Ambidentate ligands are the ligands with two or more different sites, only one of which being attached to single metal atom at a given time. Examples are :

$M{-}NO_2$	M–CN	M–SCN
M–ONO	M–NC	M–NCS

LIGAND ISOMERISM

Many ligands are themselves capable of existing in isomeric states. *When such ligands are associated to complexes, the complexes are isomers of each other. Such isomers are known as ligand isomers and the phenomenon is known as ligand isomerism.* For example, ($Coen_2Cl$ Chloroaniline)$^+$ exists in three isomers because chloroamine may be *o–,m–,p–*.

Diamino propane is another ligand which can exist both as 1,2–diamino propane (p_n) and 1,3–diamino propane (t_n)

$$\underset{\underset{\text{(p}_n\text{)}}{\text{1, 2–Diaminopropane}}}{\begin{array}{lll} CH_2 & - CH & - CH_3 \\ | & \;\,| & \\ NH_2 & \;\,NH_2 & \end{array}} \qquad \underset{\underset{\text{(t}_n\text{)}}{\text{1, 3–Diaminopropane}}}{\begin{array}{lll} CH_2 & - CH_2 & - CH_2 \\ | & \;\,| & \\ NH_2 & \;\,NH_2 & \end{array}}$$

When p_n and t_n are associated into complexes, the so obtained complexes are isomer of each other. One example of isomeric complexes having this ligand is $[Co(Pn)_2\ Cl_2]^+$ and $[Co(t_n)^2\ Cl_2]^+$ ions.

Valence Isomerism : This type of isomerism occurs in polynuclear complexes. *They have the same molecular composition but differentiate from each ether with respect to the bonding present in them.* The typical example of this type is given below :

Nuclear Co–ordination Polymerisation Isomerism : *This is present in bridged complexes and takes place due to different number of nuclei present in them.*

Electronic Isomerism : The complex $[Co(NH_3)_5NO]^{2+}$ exists in two forms. The chloride of one is black and paramagnetic while the chloride of other is pink and diamagnetic. But the structural studies of these complexes reveal that the black complex contains neutral NO and Co(II) whele pink complex contains NO^- and Co(III).

COORDINATION ISOMERISM

If both cation and union are complex, there may be an exchange of ligands between the two giving rise to isomers known as coordination isomers and the phenomenon is known as coordination isomerism. Some examples showing coordination isomerism are as follows :

(a) *The first example is* $[Cu(NH_3)_4]^{2+}$ $[PtCl_4]^{2-}$

and $[Pt(NH_3)_4]^{2+}$ $[CuCl_4]^{2-}$

In the above example, both compounds have the same chemical composition but are made up of different complex ions. The first complex has the cation $[Cu(NH_3)_4]^{2+}$ and anion $[PtCl_4]^{-2}$ while the second compound has the cation $[Pt(NH_3)_4]^{2+}$ and the anion $[CuCl_4]^{2-}$. Thus copper is in the cation of the first complex compound and in the anion of the second while for platinum it is just the reverse. Hence the above two compounds are called coordination isomers.

Another example is : $[Co(NH_3)_6][Cr(CN)_6]$

and $[Cr(NH_3)_6]$ $[Co(CN)_6]$.

In the above example cobalt occurs in the cation of the first and in the anion of the second, whereas for chromium the situation is reverse. Other examples of coordination isomerism are :

(i) $[Co(NH_3)_6]$ $[Cr(C_2O_4)_3]$ and $[Cr(NH_3)_6]$ $[Co(C_2O_4)_3]$

(ii) $[Co(en)_3]$ $[Cr(CN)_6]$ and $[Cr(en)_3]$ $[Co(CN)_6]$.

(b) Another form of the same type of isomerism may occur when the same metal appears in both cation and anion of two compounds in which there is a different distribution of donor components.

Compounds exemplifying this arrangement are :

(i) $[Cr(NH_3)_6]$ $[Cr(CNS)_6]$

and $[Cr(NH_3)_4\ (CNS)_2]$ $[Cr(NH_3)_2(CNS)_4]$

(ii) $[Cr(NH_3)_6]$ $[Cr(C_2O_4)_3]$

and $[Cr(NH_3)_4(C_2O_4)]$ $[Cr(NH_3)_2(C_2O_4)_2]$

(c) *Sometimes the same metal which appears in both cation and anion may have different oxidation states, e.g.,*

$$\overset{II}{[Pt(NH_3)_4]}\ \overset{IV}{[PtCl_6]} \text{ and } \overset{IV}{[PtNH_3)_4Cl_2]}\ \overset{II}{[PtCl_4]}.$$

POLYNUCLEAR COMPLEXES

Complexes having more than one central ion are termed polynuclear complexes. The formation of polynuclear complexes is due to the fact that a ligand bonded to a metal ion still has the ability to donate another pair of electrons.

This is the case when the co-ordinated donor atom still has unshared pairs of electrons, or the number of donor groups in the ligand exceeds the maximum co-ordination number of the central ion or the steric arrangement of the donor atoms of the multidentate ligand does not make it possible for all the donor atoms to be co-ordinated to the same central ion.

CLASSIFICATION OF THE COMPLEXES

In polynuclear complexes, bridging groups are situated between the central ions. The bridging groups, may be monatomic, oligoatomic or polyatomic.

On the basis of bridging groups the polynuclear complexes may be divided into two groups.

(*a*) *Polynuclear complexes with monoatomic bridges* : In complexes containing a mono-atomic bridge, the same donor atom is coordinated to two central ions. Examples are :

$$[Ag - I - Ag]^+ \qquad \left(\begin{matrix} & H & \\ & | & \\ Zn - & O & - Zn \end{matrix}\right)^{3+}$$

(a) (b)

Only atoms having more than one unshared pair of electrons can function as monoatimic bridges. This is a necessity but sufficient condition. For example, the water molecule cannot be a bridging ligand, although there are two pairs of unshared pairs of electrons on its oxygen atom. The following ligands may function as monoatomic bridges :

$$F^-, Cl^-, Br^-, I^-, O^{2-}\ HO^-, S^{2-}, HS^-\ Se^{2-}, Te^{2-}$$

There are some other ligands, which act as bridging ligands in the solid state, such as P^{3-}, As^{3-} and Sb^{3-}.

The formation of polynuclear complexes with monoatomic bridges is very common in the hydrolysis of metal ions. These complexes contain O^{2-} and OH^- bridges.

(*b*) *Polynuclear complexes with polyatomic bridges.* In the oligometric or polyatomic bridges the donor atoms coordinated to the atomic ions are neighbours or separated by only one further atom.

For example, some polyatomic bridges are :

Me – CN – Me – O – O – Me

Cyanide bridge Peroxo group

At present there are only a few known examples of complexes of the type involving polyatomic bridges.

The metal salts of different cyano complexes can be regarded as complexes of this type. These complexes are in fact polynuclear complexes but their small solubility makes equilibrium studies difficult.

Alternative Classification. Polynuclear complexes may be classified as follows :

(i) ***Complexes with –OH–Oi–O–bridges.*** *These complexes are formed in the hydrolysis reactions of* Th^{4+}. *In these complexes, the OH groups are shared by* Th^{4+} *ions.*

$$2Th4^{+} + 2H_2O \rightleftharpoons [Th_2(OH_2]^{6+} + 2H^{+}$$

$$2Th^{4+} + H_2O \rightleftharpoons [Th_2(OH)]^{7+} + H^{+}$$

The binuclear complex of Co (III) is formed during the hydrolytic reaction of Co (III) compound in solution–*e.g.*,

Among the many polynuclear complexes of cobalt (III) containing OH bridges, tetranuclear complex is very important. Its structure is as follows :

The complex is prepared by heating $[Cr(en)_2 (OH) (H_2O)]Cl_2$. The oxidation of Cr^{2+} by two electron–oxidising agents such as HClO or H_2O_2 gives dinuclear complex whose formula is probably $(H_2O)_5$ Cr – O – Cr $(H_2O)_5]^{4+}$.

The two electron oxidation takes place in shapes with Cr^{4+} intervening.

$$\underset{(+\,2)}{Cr^{2+}} + ClO^{+} \longrightarrow \underset{(+\,4)}{\text{‘}CrO^{2+}} + Cl^{-}$$

$$CrO^{2+} + Cr^{2+} \longrightarrow \underset{\text{dinuclear species}}{[Cr^{3+} - O - Cr^{3+}]^{4+}}$$

General Discussion : Some facts about polynuclear complexes are :

1. The stability of the polynuclear complexes increases with increasing number of functional groups.

2. If the central metal ions are identical, the complex may be termed *polynuclear,* otherwise *heteropolynuclear.* The term *homonuclear* is sometimes used when the nuclearity of all the complexes in the equilibrium system is the same.
3. Polynuclear complexes, usually binuclear, are formed in the reaction of metal complexes with another central ion. If the structure of the ligand makes it possible, a binuclear complex of a whole series of complexes is formed as intermediate.

IMPORTANCE OF POLYNUCLEAR COMPLEXES

1. The polynuclear complexes may play a role in the formation of complexes. It is well known that the rate of formation of the EDTA complex of chromium (III) increases with increasing pH. It was observed further that if the initial pH of the solution is greater than two, the rate decreases, if aged solutions are used instead of freshly prepared ones. The reason is that above pH 2, a binuclear species is formed slowly and this complex reacts much more slowly with EDTA than a mononuclear does.
2. The formation of polynuclear complexes may be important in the catalytic reaction too. The source of the catalysis of reaction of halide complexes with metal ion is the affinity of these catalysis ions for the bound halide. It is likely that binuclear complexes are intermediate in these reactions.
3. Binuclear complexes themselves may be catalysts of certain reactions. In the presence of excess of EDTA, the catalytic effect of iron (III) on the decomposition of hydrogen peroxide is completely inhibited. However, if iron (III) is in excess, the activity is even greater than in the absence of EDTA. The probable explanation is that in a binuclear complex two iron (III) ions exert a concerted action on the same peroxide molecule.

INNER COMPLEXES

These are the non-electrolyte complexes which are formed when the cationic charge is compensated by the number of negatively charged ligands and the remaining co-ordination positions are occupied by the neutral ligands.

It means that an inner complex is a non-ionic chelated structures formed by the union of a metal ion with a bidentate group which possesses a charge of minus one. Thus, the co-ordination number of the metal ion must be twice its ionic charge; this is always not the case.

Examples:

Inner complexes containing beryllium, aluminium, cobalt (III), iron (III) and chromium (III) are common : those containing cobalt (II) or iron (II) are rare because the usual co-ordination number of these ions is six; they could form inner complexes by union with a tridentate ligand of uni-negative charge. Some examples of inner complexes are :

1. Many of them are insoluble in water, but may be extracted into organic solvents immiscible with water. Their exreactibility is often a function of pH of the aqueous phase, so that selective extraction and subsequent return to a new aqueous phase are possible.
2. The formation of inner complexes is sometimes accompanied by pronounced colour changes, which permit colorimetric measurements.
3. Most of the inner complexes have low melting point and are even volatile.

Applications of Inner Complexes

Main applications of inner complexes are outlined below :

1. Due to the electrical neutrality of inner complexes, these can be extracted into a water-immiscible organic phase. This process is very useful in both analytical and industrial separations. For example, α-theonyltifluoroacetone (TTA) is used for the separation of hafnium from zirconium. This process is of great importance in nuclear reactor. Beryllium and scandium, also valuable nuclear engineering metals, have also been refined by solvent extraction in the presence of TTA.
2. The inner complexes are being used in analytical chemistry for the estimation of ions. For example :
   (a) The 8–hydroxyquinoline (oxine) co-ordinates with numerous metallic cations to give inner complexes. For each ion there is a definite pH range in which inner complex is formed (Table 1).

Table 1

Cation	Cu^{2+}	Fe^{2+}	Al^{3+}	Th^{4+}	Mn^{2+}	Mg^{2+}
pH range	> 2.7	2.8–11.2	4.2–9.8	4.8–8.8	5.9–10	> 8.2

(b) Compounds like dialkyl glyoxime (I) are capable of forming inner complexes (II). But o–quinone dioxime (III) will not form this type of compound. However, this compound on reduction yields corresponding cyclohexane derivative which forms a stable inner complex.

STABILITY CONSTANT

When ammonia is added to an aqueous solution of, say, nickel sulphate, $NiSO_4$, there is a noticeable colour change because some of the $[Ni(H_2O)_6]^{2+}$ ions are converted into $[Ni(NH_3)_6]^+$ ions. The reaction taking place is a replacement reaction involving ligands :

$$[Ni(H_2O)_6]^{2+} + 6NH_3 \rightleftharpoons [Ni(NH_3)_6]^{2+} + 6H_2O$$

The equilibrium constant (K) for this reaction is,

$$[K = \frac{[Ni(NH_3)_6]^{2+}\,[H_2O]^6}{[Ni(H_2O)_6]^{2+}\,[NH_3]^6}$$

A high value of equilibrium constant would mean that the $Ni\,(NH_3)]^{2+}$ *is more stable.* In aqueous solution, the concentration of water can be regarded as constant so that the value of equilibrium constant K is

$$K = \frac{[Ni(NH_3)_6]^{2+}}{[Ni(H_2O)_6]^{3+}\,[NH_3]^6}$$

K is known as the stability constant for the $[Ni(NH_3)_6]$ ion. *The higher the stability constant for a complex ion, the greater its stability.*

Alternatively, 1/K *values are sometimes called* instability constants. *A high value for an instability constant means an unstable complex ion.* Generally, stability constants are after quoted as $\log_{10}$ K and some typical value are given below.

$[Co(NH_3)_6]^{2+}$	5.11	$[Ni(NH_3)_6]^{2+}$	7.99
$[Zn(NH_3)_4]^{2+}$	9.11	$[Cu(NH_3)_4]^{2+}$	12.60
$[Fe(CN)_6]^{4-}$	8.30	$[Fe(CN)_6)]^{3-}$	01.00

Stepwise Stability Constants

According to I. Bjerrum (1941), the formation of a complex in solution proceeds by the stepwise addition of ligand to a metal. Thus, when a metal ion forms complexes with a ligand, a number of successive equilibria can be formulated. These are :

$$M + L \rightleftharpoons ML \; ; K_1 = \frac{[ML]}{[M][L]}$$

$$M + L \rightleftharpoons ML_2, K_2 = \frac{[ML_2]}{[M][L]}$$

$$ML_2 + L \rightleftharpoons ML_3 \; ; K_3 = \frac{[ML_3]}{[ML_2][L]}$$

... ...

$$ML_{N-1} + L \rightleftharpoons ML_N \; ; KN = \frac{[ML_N]}{[ML_{N-1}][L]}$$

Where N represents the maximum co-ordination number of the metal ion for the ligand L.

K_1, K_2, K_3, ... K_N are the **stepwise formation or stability** constants and are related to the total or overall stability constant K, by the equation.

$$K_1 = K_1 \times K_2 \times K_3 \times \ldots\ldots K_N$$

for the total reaction $M + NL \longrightarrow ML_N$

$$K_t = \frac{[ML_N]}{[M][L]^{N\cdot}}$$

Relation Between Entropy and Stability Constant

The standard free energy charge ΔF° is related to the equilibrium constant by the relation,

$$\Delta F^\circ = - RT \text{ in } K_t \qquad \ldots (A)$$

where R is the gas constant and T is the absolute temperature. Also,

$$\Delta F^\circ = \Delta H^\circ \; T\Delta S^\circ \qquad ..(B)$$

From these two relations (A) and (B), we get

$$\Delta H^\circ - T \; A \; S^\circ = - RT \ln K_t$$

From the above equation, it follows that complex formation is favoured by *negative enthalpy changes and positive entropy changes.*

Methods for Determination of Stability constants and Formulae of Complexes

The method of continuous variation (Job's Method) : This method is used for the determination of formula of a complex and dissociation constants to complexes of low solubility.

Principle : The method of continuous variation makes use of any measurable additive property, provided that this property has different values of for the various species present in solution of a metal (M) and ligand (L). This method is mainly used for such solutions where only one complex is formed.

Procedure : Continuous variation method generally utilises optical density as an additive property. Its various steps are :

(i) *Make up a series of mixtures containing a total of 10 ml of metal ion and ligand indifferent proportions, e.g.,* 0–10, 1 – 9,... 9 – 1, 10–0. It means that the sum of total analytical concentration C of complexing agent C_x and metal ion C_M is held constant and only that ratios are varied :

$$C_x + C_M = C \quad ...(1)$$

(ii) Then, determine the optical densities of solutions prepared as in step (i) by means of a spectrophotometer at such a wave length of light that complex absorbs light strongly whereas the metal ion and ligand do not.

(iii) Draw a graph between mole fraction of ligand X ($X = C_x/C$) and optical density. Extrapolate the legs of the curve until they cross. At the point of intersection the mole fraction gives the formula of a complex MX_n, *i.e.*,

$$n = \frac{C_X}{C_M} \quad ...(2)$$

Equation (1) may be written as :

$$\frac{C_X}{C} + \frac{C_M}{C} = \frac{C}{C}$$

But

$$\frac{C_X}{C} = X \quad ...(3)$$

or

$$X + \frac{C_M}{C} = 1 \text{ or } \frac{C_M}{C} = 1 - X \quad ...(4)$$

On dividing equation (3) by (4), we get

or $$\frac{C_x}{C} = \frac{X}{1-X}$$

or $$n = \frac{X}{1-X} \qquad \left[\because n\,\frac{C_X}{C_M} \text{ from (2)}\right]$$

From the value of n, the formula of a complex MX_n can be determined.

Limitations of the Method

(i) The method of continuous variation applies provided that not more than one complex is formed under the given experimental conditions.

(ii) This method is only applicable when there is no volume change on mixing the solutions of the metal ions and ligands.

THEORY OF LOGARITHMIC METHOD

The limiting logarithmic method consists in measuring the optical densities of solutions of metal (M) ion and ligand (L) in which the concentrations of the two species are varied in turn at a constant total concentration. It is assumed that Beer's law is applied, *i.e.,* the optical density at the absorption peak is proportional to the concentration of the complex.

For the reaction

$$mM + lL \longrightarrow M_mL_l$$

The stability constant. K is given by

$$K = \frac{[M_mL_l]}{[M]^m\,[L]^l}$$

On taking logarithms, we get,

$$\log\,[M_mL_l] = m\log[M] + l\,\log[L] + \log K \qquad ...(1)$$

The value of m, l and K may be found by plotting the logarithms of the optical densities of two sets of solutions versus composition.

(i) *One with constant metal and variable ligand concentration, and*

(ii) *The other with constant ligand and variable metal concentration*
The slope of the curve in case (i) yields "l " while in case of (ii) "m" is obtained. From these values of "l" *and "m"* and equation (5), values of K and the composition of the complex can be calculated.

Limitations

(i) *The method is only suitable at low concentration.*

(ii) *The value of stability constant obtained by the method is higher than that given in the literature.*

Molar Ratio Method : This method is vary simple as compared to the continuous variation method. The difference between two methods is that in the molar ratio method, the total analytical concentration of metal (or ligand) is held constant rather than the sum of ligand and metal concentration.

Optical density is measured at such a wavelength of light at which complex absorbs strongly whereas the ligands and metal ion do not. Then draw a graph between absorbance, *i.e.,* optical density versus ligand to the metal concentrations $\left(\frac{C_x}{C_m}\right)$. Extrapolate the straight line C portions until they cross. At this point, the value of $R = \frac{C_x}{C_m}$ gives the ratio of ligand to the metal ion in the complex

Matching Absorbance Method : This method is quite useful for the determination of stability constants. This method involves the following steps :

(i) The absorbance of two solutions having different metal to ligand rations are measured at a wavelength where only the complex absorbs strongly, and the ligand and metal ion do not.

(ii) Then dilute the solution of higher absorbance to such an extent until the absorbance of the two solutions becomes the same.

(ii) Let the concentration of metal and ligand in solution (1) be C_{m1} and C_{X1} respectively. For the solution (2) C_{m2} and C_{x2} are the corresponding quantities.

(iv) For 1 : 1 complex, we have

$$[M_1] = C_{m1} - [MX]_1$$

$$[X_1] = C_{x1} - [MX]_1$$

$$[M_2] = C_{m2} - [MX]_2$$

$$[X_2] = C_{x2} - [MX]_2$$

When the absorbance of the two solutions becomes equal then $[MX]_1 = [MX]_2 = Cs$ where Cs is some unknown concentration.

(v) The stability constant K can be calculated from the stability constant expression which contains Cs and K as unknown by eliminating Cs. Thus,

$$K = \frac{[MX_1]}{[M_1][X_1]} = \frac{[MX_2]}{[M_2][X_2]}$$

or $$\frac{Cs}{(C_{ml} - Cs)(Cs_{x1} - Cs)} = \frac{Cs}{(C_{m2} - Cs)(C_{x2} - Cs)}$$

$$Cs = \frac{(C_{m1}C_{x1}) - (C_{m2}C_{x2})}{(C_{ml} + C_{x1}) - (C_{m2}C_{x2})} \quad ...(1)$$

Thus, the stability constant can be easily determined.

Bjerrum Method : Bjerrum determined formation constants of metal complexes and chelate compounds in aqueous solutions. Bjerrum described the stepwise formation of series of metal complexes of the type ML, ML_2 ... ML_n which are defined by equations (2) to (4).

$$M + L \longrightarrow ML \; K_1 = \frac{[ML]}{[M][L]} \quad ...(2)$$

$$ML + L \longrightarrow ML_2 \; K_2 = \frac{[ML_2]}{[M][L]} \quad ...(3)$$

$$ML_{n-1} + L \longrightarrow ML_n \; K_n = \frac{[ML_N]}{[ML_{n-1}][L]} \quad ...(4)$$

Equation (2) can be written as

$$[ML] = K_1 [M] [L] \quad ...(5)$$

Equation (3) can be written as

$$[ML_2] = K_1 [ML] [L] \quad ...(6)$$

Substituting the value of [ml from equation (5) in (6),

$$[ML_2] = K_1K_2[L_2] [M]$$

Similarly, equation (4) can be written as

$$[ML_n] = K_1, K_2, K_3, ...K_n [L]_n [M] \quad ...(7)$$

Bjerrum introduced a function $\overline{n}$, which is defined as *the average number of ligand molecules bound per mole of metal.* It may be expressed mathematically as :

$$\overline{n} = \frac{[ML] + 2[ML_2] + 3[ML_3] + ... + n[ML_n]}{[M] + [ML] + [ML_2] + [ML_3] + ... + [ML_n]} \quad ...(8)$$

Substituting eqs. (5), (6) and (7) in (8) we get

$$\overline{n} = \frac{K_1[M][L] + 2K_1 K_2[M][L]^2 + nK_1 K_2 K_3 ... + K_n[M][L]^n}{[M] + K_1[M][L] + K_1 K_2[M][L]^2 + ... + K_1 K_2 K_3 ... K_n[M][L]^n}$$

Cancelling 'M' throughout, we get

$$\overline{n} = \frac{K_1[M][L] + 2K_1 K_2[L]^2 + ... + nK_1 K_2 K_3 ... K_n[L]^n}{1 + K_1[L] + K_1 K_2[L]^2 + ... + K_1 K_2 K_3 + ... K_n[L]^n}$$

Equation (9) is the so called Bjerrum formation function. If the concentration of unbond ligand (A) can be calculated experimentally, $\overline{n}$ can be calculated form the equation :

$$\overline{n} = \frac{L_t - [L]}{M_t}$$

where L_t and M_t denote the total concentration of ligand and metal respectively.

Solution of equation (14) for the known values of n and corresponding (L) values yields values of the formation constants, *i.e.,* K_1, K_2, K_3...K_n.

Let us illustrate the Bjerrum method by calculating the formation constants for the complex between $CuSO_4$ and 5-sulphosalicylic acid (H_3L).

Procedure

Titrate 100 ml of each solution *i.e.,* $CuSO_4$ and 5-sulphosalicylic acid with standard NaOH solution at 25°C, noting the pH values after addition of the titrant base NaOH .

Plot the pH *versus ml of base* curves and note the point of inflexion of the curve corresponding to the mixed solution. Read off the horizontal distance between the curves. Thus find values of $\overline{n}$ and plot the formation curve of $\overline{n}$ against $-\log [L^{3-}]$, the values of the latter having been found at various pH from the relation:

$$[L^{3-}] = \frac{[HL_3]_{total} - [CuL^-] - 2[CuL_2^{4-}]}{\dfrac{[H^+]^2}{[K_{II}\,K_{III}]} + \dfrac{[H^+]}{[K_{III}]} + 1}$$

where K_{II} and K_{III} are the second and third dissociation constants of HL_3 which are 3.23×10^{-3} and 1.81×10^{-3} respectively.

The values of K_1 and K_2 found from the formation curve at $\bar{n}$ = 0.5 and K_1 = 1.5 are $K_1 = 2.2 \times 10^9$ and $K_2 = 6.3 \times 10^6$.

Polarographic Method : This method is based on the fact that the reduction potentials of metal ions are usually shifted to the more negative values as a result of complex formation. Consider the reaction.

$$ML_m + ne + Hg \xrightarrow{k} MHg + m\,L \qquad ...(9)$$

which may be imagined to occur in two steps :

$$ML_m \xrightarrow{k_1} M + m\,L \qquad ...(10)$$

$$M + ne + Hg \xrightarrow{k_2} MgH + m\,L \qquad ...(11)$$

where k, k_1 and k_2 correspond to respective formation constants of the above reaction ; M is the metal, L the ligands, ML_m is the complex compound and *n* is the number of electrons involved in the reduction. *Lingane* derived from the above reactions a relation connecting the potential at the dropping mercury cathode, E, the half wave potential $E_{0.5}$., the current *i* and the limiting current *i* (the maximum of the polarographic wave) viz,

$$E = E_{0.5} - \frac{0.0591}{n} \log \frac{i}{i_{d-i}} \qquad ...(12)$$

Plot of E against log (i/i_d–i) will, of course, be a straight line for a reversible potential, the slope of the line being 0.059/n where *n* is the number of electrons transferred in the reduction.

By considering equations (17), (18) and (19) Lingane found that at 25°C

$$E_{0.5} = E^\circ - \frac{0.0591}{n} \log k - \frac{0.0591}{n} \log \frac{k_1}{k_2} - m \frac{0.0591}{n} \log [L]$$

where E° is the standard potential of the metal amalgam electrode, the ratio k_1/k_2 depends upon experimental conditions and electrode characteristics, *m* is the number of co-ordinated ligands, [L] is the ligand concentration. Hence, the slope of line obtained by plotting $E_{0.5}$ against log [L] is $-\frac{0.0591m}{n}$ from which *m* can be found as the value of n is known. On the other hand, the stability or formation constant *k* of reaction (9) is found from the equation :

$$E_{0.5} - E_{(0.5)s} = -\frac{0.0591}{n} \log k - m\frac{0.0591}{n} \log[L] \quad ...(21)$$

where $E_{(0.5)s}$ is the half wave potential in the absence of ligands.

The plot of the left hand side of equation (21) against L yields a straight line, the intercept of which on the co-ordinate axis gives $\left(\frac{0.0591}{n}\right)$ log k and hence k can be determined. It is also clear that the slope of the line is $-\left(\frac{m \times 0.0591}{n}\right)$ so that *m* can be calculated.

SOLUBILITY METHOD

If the metal ion forms a relatively insoluble salt whose solubility product is known and a complex which is rather more soluble, the stability constant of the complex may be determined by measuring the increased solubility of the salt caused by the presence of the ligand. Consider the reaction of silver acetate with acetate ion. If a sodium acetate solution of known high concentration is equilibrated at constant, temperature with an excess of silver acetate, the equilibrium given in eq. (I) is satisfied.

$$AgOAc \rightleftharpoons Ag^+ + OAc^-$$

$$K_{sp} \rightleftharpoons [Ag^+)[OAc^-]$$

$$Ag^+ + 2OAc^- \rightleftharpoons Ag(OAc)_2^-$$

$$\beta_2 = \frac{[Ag(OAc)_2^-]}{[Ag^+][(OAc^-]^2} \quad ...(1)$$

After equilibration, the excess solid silver acetate is removed by filtration and the solution is analysed for total silver. Then, the following relationship holds.

$$C_{Ag}{}^+ = [Ag(OAc_2)^-] \quad ...(2)$$

where $C_{Ag}{}^+$ is the silver solution concentration. From the solubility product expression, we get, $[Ag^+] = K_{sp}/[OAc^-]$, and this then becomes

$$[Ag^+] = K_{sp}/_{CNaOAc} \quad ...(3)$$

because of the large excess of sodium acetate. Substituting equations (2) and (3) in the formation constant expression which gives

$$\beta_2 = (CAg^+ - K_{sp}/C_{NaOAc})/[(K_{sp}/C_{NaOAc}\ C_{NaOAc})^2] \quad ...(4)$$

Each of these quantities is known, and the overall stability constant β_2 may be calculated.

Electromotive Force Method : A metal in contact with a solution of its ions gives a potential, E, which is governed by equation (1)

$$E = E° - (RT/n\ F)\ \ln([M^{n+}]/[M]\ \text{solid})$$

If two different cells of this type are connected, the voltage is the difference of the two cell potentials and can be measured with high precision.

If two cells are identical *i.e.,* both have same metal and the same metal ion concentration, the measured voltage will be zero. If a ligand be added to cell 1, the complex equilibrium will cause a change in the free metal ion

$$M^{2+} + L \rightleftharpoons ML^{2+} \quad ...(5)$$

concentration and a net voltage will be observed. The observed voltage is given by eq. (3) or (4).

$$E_{obs} = E_1 - E_2$$

$$= -(RT/n\ F)\ \ln[M^{2+}]1/[M]_{solid} - \ln([M^{2+}]_2/[M]_{solid})$$

$$E_{obs} = -(RT/n\ F)\ \ln([M^{2+}]_1/[M^{2+}]_2)$$

where T is the absolute temperature, R is the gas constant ; F is the Faraday (96500 coulombs); and n is the number of electrons transferred, in this case ($M° \longrightarrow M^{2+}$) equal to 2, Thus, the observed voltage is directly related to the free metal ion concentrations in the cells. In practice successive portions of ligands are added to the metal ion solution in one cell, and the corresponding voltages measured. From the known total concentration of the metal ion, ligand, and uncomplexed metal ion, the stability constant may be evaluated.

STABILISATION OF THE OXIDATION STATES BY COMPLEX FORMATION

It has been observed that in a large number of cases the oxidation states of an element which are otherwise unstable or less stable than the other oxidation states are *Stabilised* by co-ordination with suitable complexing reagents.

Many oxidation states which are not normal for metal can be stabilised thermodynamically and sometimes kinetically, by co-ordination with a suitable ligand.

An early example of this phenomenon occurred with the +2 and +3 oxidation states of cobalt. In strong $HClO_4$ media both Co^{2+} and Co^{3+} can be prepared and it is found that the +2 state is stable but the +3 state is not. For example :

(i) Co^{3+} rapidly and completely oxidises Fe^{2+}, SO_3^{2-} and even Cl^- producing CO^{2+}

(ii) In the absence of reducing agents, Co^{3+} slowly oxidises the solvent water, producing oxygen.

(iii) In the presence of ethylene diamine, $[Co(en)_2]^{3+}$ is both kinetically and thermodynamically more stable than $[Co(en)_3]^{2+}$. The former will not oxidise most reducing agents, while the latter will reduce Fe and many other oxidising agents.

A qualitative explanation of the effect based on the crystal field theory will now be presented.

Application of Crystal Field Theory

Neglecting the electron pairing energy and the electron ionisation energy, since they will be essentially constant for both ligands, the crystal field stabilisation energy (CFSE) for the +2 ion is $[5(.0.4) + 2 (+ 0.6)]\ \Delta_0 = -0.8\Delta_0$ and for the +3 ion is $[6\ (.0.4) + 0\ (+ 0.6)]\ \Delta_0 = -2.4\ \Delta_0$.

The large negative value of the CESE for the state at any Δ_0 value suggests that the +3 state is more sensitive to the ligand field and as Δ_0 increases ($H_2O \rightarrow$ en) the + 3 state would show a larger CFSE change than the + 2 state.

Therefore, the + 3 oxidation state of cobalt would be stabilised (compared to Co^{+2}) with ethylene diamine ligands.

Many other examples can be shown where the oxidation potential is drastically modified in the presence of a co-ordinating agent. From this it can be seen that F~ and CN^- stabilise the +3 state while dipy, and *o*–phen, stabilise the + 2 state when compared with water.

Table 2

System			*Volts*
$[Fe(H_2O)_x]^{2+}$	$\longrightarrow$	$[Fe(H_2O)_x]^{3+} + e^-$	– 077
$[FeFe_6]^{4-}$	$\longrightarrow$	$[FeFe_6]^{3-} + e^-$	– 0.40
$[Fe(CN_6]^{4-}$	$\longrightarrow$	$[Fe(CN)_6]^{3-} + e^-$	– 0.36
$[Fe(dipy)_3]^{2+}$	$\longrightarrow$	$[Fe(dipy)_3]^{3+} + e^-$	– 1.10

Thermodynamically it appears that :

(a) low oxidation states will be favoured with ligands having reducing power or with reducing solvents,

(b) Basic conditions will favour high oxidation states and acid conditions low,

(c) The electronic configurations of the oxidation states coupled with the ligand field strength may stabilise either the upper or lower state.

Kinetic stabilisation of a particular oxidation state may be affected through (a) complex insolubility and (b) the use of bulky ligands which prevent otherwise favourable reaction routes through steric hindrance.

We will now discuss some uncommon oxidation states and show how these are stabilised through complex formation,

Stabilisation of Ag (II) and Ag (III) : Silver can exist in the 1 +, 2 + and 3 + oxidation states, but only the first is normally observed.

Stabilisation of Ag (II) : Solution of the argentic ion Ag^{2+} in perchloric and nitric acids may be prepared by oxidation of Ag^+ with ozone. Kinetic studies have suggested that this oxidation occurs by a mechanism involving Ag^+.

$$Ag^+ + O_3 \longrightarrow AgO^+ + O_2$$

$$AgO^+ + Ag^+ + 2H^+ \longrightarrow 2Ag^{2+}\ H_2O$$

The potentials for the AgO^{2+}/Ag^+ couple in 4M $HClO_4$ show that Ag^{2+} is a powerful oxidising agent. In the presence of oxidising agents and a ligand capable of double bonding such as py, dipy or O–phen, Ag^{2+} complexes are formed which are remarkably stable. The complexes $[Ag(py)_4]^{2+}$ and $[Ag\ (dipy)_2]^{2+}$ are coloured (red to brown) having unpaired electron, and have the ligands arranged in a plane. Similar compounds can be prepared with tripyridyl and picolinic acid.

All complexes of Ag^{2+} have magnetic moments of 1.75-2.2 B.M. in accord with expectation for the d^9 configuration; spectra are also consistent with a planar structure.

Stablisation of Ag (III) : Few complexes of Ag^{3+} complexes are known. Ag^{3+} complex of remarkable stability is the one with ethylene-dibiguanide (as shown below) which is obtained as the red sulphate when Ag_2SO_4 is treated with aqueous potassium peroxidisulphate in the presence of ethylenedibiguanidinium sulphate.

Stabilisation of Cu(I) and Cu(III) : Copper can exist in 1 +, 2 + and 3 + oxidation states, but only 2 + is normally observed.

Stabilisation of Cu (I) : The relative stabilities of cuprous and cupric states are indicated by the following potential data :

$$Cu^+ + e^- \longrightarrow Cu \qquad Eo = 0.52\ V$$

$$Cu^{2+} + e^- \longrightarrow Cu^+ \qquad Eo = 0.152V$$

$$\text{whereas}\quad Cu + Cu^{2+} \longrightarrow 2Cu^+ \qquad E_0 = -\ 0.37\ V :$$

$$k = [Cu^{2+}]/[\ Cu^+] = \sim 10^6$$

In other words Cu^+ ions under these conditions, are more stable than Cu^+ ions : the equilibrium constant for the above reaction is 10^6. In the reaction, one Cu^+ ion is converted into a Cu atom (with a lower oxidation state) whilst the other Cu^+ ion is converted into a Cu^{2+} ion (with a higher oxidation state); a change of this nature is known as *disproportionation.* Cuprous (I) gets stable Used by forming complexes with non–π bonding as well as *p*π and *d*π bonding ligands.

Complexes of Cu^+ ions, *e.g.* $Cu(NH_3)_2^-$ and $Cu(NH_3)_4^+$, are, however, stable in aqueous solution, as are insoluble copper (I) compounds, for example, copper (I) chloride, iodide and cyanide.

With ammonia, halide ions and monodentate ligands, the cuprous complexes usually have the highest stability for $n = 2$ *e.g.* $[Cu(NH_3)_2]^+$. However, in general the complexes from $n = 1$ to n = 4 can exist under appropriate conditions. The cyanide complex, $K[Cu(CN)_2]$ has been shown to have unusual spiral polymeric. Structure where the copper atom has co-ordination number 3, the nitrogen and carbon atoms bcund to copper are almost coplanar with the metal atom

Stabilisation of Cu (III) : There is a good evidence that Cu^{3+} can occur in crystalline compounds and in complexes. Cu^{3+} can be stabilised by forming a complex K_3CuF_6 which is obtained by treatment of a mixture of potassium and cupric fluorides with fluorine. K_3CuF_6 is a pale green crystalline solid and is paramagnetic with $\mu = 2.8$ B.M.

Stabilisation of Co (III) and Co (I) : The unusual oxidation states of cobalt are 3 + and 1 +. However, these can be stabilised by forming complexes.

Stabilisation of Co (III) : Simple cobalt (III) salts are difficult to make, and unstable when made, because it is difficult to oxidise Co^{2+} to Co^{3+} even using electrolytic methods. However, Co (III) gets stable by forming complex like $[Co(NH_3)_6]^{3+}$ and $[Co(CN)_6]^{3-}$ ions which are very stable. So much so, that even air will oxidise Co^{2+} to Co^{3+} in the presence of cyanide ions or ammonia and the resulting cobalt (III) complexes are much more stable than the corresponding cobalt (II) ones.

This may be due to the fact that cobalt (III) complexes involving d^2sp^3 hybridisation would have the 3*d*–,4*s*– and 4*p*– orbitals just filled, where as the cobalt (II) complexes would have one electron in the higher 5s level if they were to form d^2sp^3 hybrid bonds.

Stabilisation of Cobalt (I) : The oxidation state can be stabilised by forming octahedral and square pyramidal complexes. For example, cobalt carbonyl reacts with isonitrilies, disproportionating to Co (I) and Co(–I).

$$Co_2(CO)_8 + 5RNC \longrightarrow [Co(CNR)_5] + [Co(CO)_4] + 4CO$$

There are several cases in which Co(II) complexes add NO to form substances conventionally formulated as $NO^+ - Co(I)$ complexes. Example is $Co(S_2CNEt_2)_2$ where a square pyramidal structure has been established by X–ray work.

Stabilisation of Fe (II) and Fe (III), Fe (IV) and Fe (VI) : Iron can exist in the +2, +3, +4 and +6 oxidation states but only the first two are normally observed. However, all the states can exist if they form complexes with suitable reagents.

Stabilisation of Fe (II) and Fe (III) : Many simple iron (II) compounds are readily oxidised, sometimes on exposure to air. However, the complexes formed by iron (II) ions with phenanthroline are much more stable. Ortho-phenanthroline iron (II) sulphate (which is orange red in colour) is oxidised to the iron (III) state (which is pale blue in solution) by strong oxidising agents.

Complexing with phenanthroline stabilises the divalent state of iron, but complex formation with cyanide ion stabilises the trivalent state. The extent of such stabilisation, when oxidation – reduction reactions are involved can be measured in terms of redox potentials, the following values illustrate the point :

$$[Fe(H_2O)_6]^{5+} \rightleftharpoons [He(M_2O)_6]^{3+} + Ie^-; E° = -0.77 \text{ volt}$$

$$[Fe(phen)_3]^{2+} \rightleftharpoons [Fe(phen)_3]^{3+} + Ie^- ; E° = -6.14 \text{ volt}$$

$$[Fe(CN)_6]^{4-} \rightleftharpoons [Fe(CN)_6]^{3-} + Ie^- ; E° = -0.3 \text{ volt}$$

Stabilisation of Fe (IV) : The best known compounds of iron in this oxidation state are Sr_2FeO_4 and Ba_2FeO_4. Both are made by oxidation of the hexahydroxoferrates with molecular oxygen at elevated temperatures.

$$M_3^{II} [Fe(OH)_6]_3 + M^{II} (OH)_2 + \frac{1}{2} O_2 \xrightarrow{800-900°C} 2M_2^{II}FeO_4 + 7H_2O$$

Another example is the salts of the ions $[Fe^{IV} (diars)_2 2X_2]^{2+}$, X = Cl, Br, which are obtained by oxidation of the corresponding $[Fe^{III} (diars)_2X_2]^+$ ions by 1.5 M nitric acid.

They have magnetic moments indicating two unpaired electrons, consistent with a spin paired $d\epsilon^4$ octahedral configuration having a large tetragonal distortion as is expected from the transpositions of the halogens.

Stabilisation of Fe (VI) : The ferrate (VI) ion, FeO_4^{2-} can be obtained by oxidising suspensions of $Fe_2O_3.nH_2O$ in concentrated alkali with chlorine or by anodic oxidation of metallic ion in concentrated alkali.

The ferrate (VI) ion is relatively stable in basic solution, but in neutral or acidic solution it decomposes according to the equation.

$$2FeO_4^{2-} + 10H^+ \longrightarrow 2Fe^{3+} + 3/2O_2 + 5H_2O$$

Stabilisation of Ni (III), Ni (I), Ni (0) and Ni (–1) : Nickel can exist in the +3, +1, 0 and –1 oxidation states, but only the state (II) is stable and most common state. However, the other states are uncommon

Stabilisation of Ni (III) : This state can be stabilised by forming complexes. For example, when certain complexes of the type Ni $(R_3P)_2X_2$ (X = Cl, Br) are treated with ClNO or BrNO, the compounds Ni $(R_3P)_3$ are obtained.

These are intensely coloured substances easily soluble in common organic liquids which are monomeric in solution and have one unpaired electron. Since their electric dipole moments are effectively zero, they are believed to have symmetrical trigonal bipyramidal structures.

Stabilisation of Ni (I), (0) and Ni (–1) : The compounds of Ni(O) are definitely the most numerous. The example of Ni(O) is $Ni(CO)_4$ in which the nickel atom is surrounded by four π-acid type ligands.

The oxidation state (–1) is represented by the carbonylate anion $[Ni_2(CO)_6]^{2-}$ of unknown structure. There is also a carbonylated anion $[NH_4(CO)_9]^{2-}$ in which the formal oxidation state of Ni is – 1/2.

The reduction of $[Ni(CN)_4]^{2-}$ with potassium metal dissolved in liquid NH_3 produces either a dark yellow precipitate $K_4[Ni(CN)_4]$, or a deep red compound $K_2Ni(CN)_3$, depending on conditions. These compounds correspond to the 0 and + 1 oxidation states respectively. Both are unstable in the presence of air. Ni(I) complexes are very rare indeed.

OPTICAL AND STERIO-ISOMERISM

This is another important type of isomerism which occurs mainly in transition metal complexes. *This type of isomerism occurs in molecules having an asymmetric atom, i.e., it can exist in two forms that are mirror images of each other, just as the right hand and the left hand.*

The two forms are identical in all respects. The only difference is that while, the one rotates plane of polarised light to the left while the other does so to the right. These are called *optical isomers*.

The isomer, which rotates the plane of polarised light towards right, *i.e.,* in clockwise direction, is said to be dextrorotatory or *d* form. The form may also be represented by placing +ve sign before its name or formula.

The isomer which rotates the plane of polarised light towards left, *i.e.,* in anticlockwise direction is termed as laevorotatory or l-from. This form is also represented by putting –ve sign before its name or formula.

The extent of rotation of the plane of polarised light by the two isomers is exactly same.

Hence, when a solution contains equal concentration of the two isomers, *i.e., d* -and l-isomers, the rotations cancel each other and the resulting solution does not rotate the plane of polarised light. Such a *d*, l-mixture which is optically inactive is called *racemic mixture.*

The d-and l-isomers of a compound are always mirror image to each other and are called enantiomorphs) Latin, enantia → opposite, morphs → forms) or enantiomers. In general for a molecule or ion to be optically active, it must not have plane of symmetry. Stereoisomerism is generally observed in complexes with coordination number 4 and 6.

Stereo-Isomerism

Werner's coordination theory predicts the directional influence of the coordinate bonds. *Isomerism arising out of the directional influence of the coordinate bonds is called stereoisomerism.* It is, by far, the most common and important of the isomerism exhibited by the coordination compounds.

Stereoisomerism is the isomerism which arises on account of the different arrangements of atoms or groups in space in a molecule (*Greek stereos = occupying space*). These different arrangements are known as stereoisomers.

Stereoisomerism in inorganic compounds relates to the central atoms having coordination number 2 to 9. Spatial arrangements for central atoms with coordination numbers 2 to 9 have observed in metal complexes but *coordination numbers* 4 *and* 6 *are more common.* Whereas, the four coordination number may give rise to either the square planar or tetrahedral complexes, the six coordination number gives rise only to octahedral complexes. Stereoisomerism is also called spaceisomerism. It is of two types:

GEOMETRICAL ISOMERISM

In metal complexes the ligands may occupy different types of positions around the central metal atom. These positions may be either next to one another (*cis*) or opposite each other (*trans*). Hence, geometrical isomerism is also called *cis-trans* isomerism. It is characterised by compounds having the same structure but different configurations. Their molecular symmetry is such that they are unable to rotate the plane of polarised light but if their structure satisfies the requirements for optical isomerism, they can also exhibit optical isomerism. The *cis-trans* isomers differ in all their physical and in many of their chemical properties. Hence, they are usually easily separated by chemical processes.

Geometrical isomerism is not shown by the complexes having coordination number 2 and 3; it has also not been found in tetrahedral (coordination number 4) complexes. In all these cases ligands occupy adjacent positions. Geometrical isomerism is of frequent occurrence in square planar (4–coordinate) and octahedral (6– coordinate) complexes.

GEOMETRICAL ISOMERISMS IN 4-COORDINATION COMPOUNDS

Square Planar Complexes

These complexes exhibit geometrical isomerism. But complexes of the type MA_4, MA_3B, MAB_3 have no geometric isomers because every possible arrangement for any of these compounds will be exactly the same. However, the following types of square planar complexes, show *cis-trans* isomerism.

Tetrahedral Complexes

Complexes with coordination number 4, mostly have square planar or tetrahedral arrangements. Geometrical isomerism is not possible in tetrahedral complexes because all the four ligands are equidistant to one another. But this isomerism is very common in square planar complexes.

$(Ma_2b_2)^{n\pm}$ Type Complexes. In these complexes, M is a metal ion and a b are monodentate ligands. Such types of complexes exist in *cis–* and trans–isomerism.

In cis–form, the two *a* groups or the two *b* groups occupy neighbour positions while in *trans* form they occupy the distant most positions. Examples of this type of complexes are $[Pt(NH_3)_2Cl_2]$ and $[Pd(NH_3)_2(NO_2)_2]$. *Cis–trans* isomers of $[Pt(NH_3)_2Cl_2]$ are shown in Fig. 3.27.

$[Ma_2bc]^{\pm}$ ***type complexes :*** *These also exist in cis and trans isomers.* For example, $[Pt\ a_2\ bc]$ type complexes exhibit *cis–trans* isomerism. Here a is a neutral ligand such as NH_3, py, H_2O and b and c are the anionic ligands such as Cl^-, Br^-, NO_2^-, SCN^- etc.

[Mabcd]$^{\pm}$ type complex : These exist in the isomeric forms. An interesting example of this type is $[Pt(NO_2)(Py)(NH_3)(NH_2OH)]^+$. This complex exists in three isomers. This complex ion possesses a plane of symmetry but it has no rotational symmetric axis because four groups are different.

Another interesting example is $[Pt(NH_3)(Py)(Cl)(Br)]^+$. This complex exists in three isomeric forms shown in Fig. 3.31.

Square planar complexes having unsymmetrical bidentate chelating ligands: These complexes are of the type $[Mab)_2]^{n\pm}$. Here M denotes the central metal ion and ab denotes an unsymmetrical bidentate ligand.

Square planar complexes having symmetrical bidentate chelating ligands: These complexes are of the type $[M(aa)_2]^{n\pm}$. These also exist in *cis*–and *trans*-isomers.

Bridged binuclear planar complexes: These exist in *cis* and *trans*-isomers as well as in the unsymmetrical isomers. For example, $[Pt(Pt_3)Cl_2]_2$ exists in *cis*-and trans and unsymmetrical isomers.

GEOMETRICAL ISOMERISM IN 6-COORDINATION COMPOUNDS

This is the most common and extensively studied coordination number. The 6–corrdinated groups can be arranged around the central metal in three forms *i.e.*,

(a) Plane hexagon

(b) Trigonal prism and

(c) Regular octahedron.

But the physical and chemical evidences have proved that the arrangement of six ligands in a 6–coordination compound is always octahedral, and the other two forms, *i.e.*, plane hexagonal and trigonal prism, are of historical interest only.

A regular octahedron has eight faces and six equivalent vertices. In an octahedral complex the metal is at the centre and the ligands are placed at the vertices.

No geometrical isomers are possible for complexes of the type Ma_6, Ma_5b and Mab_5. However, the following types of octahedral complexes show *cis–trans isomerism.*

Octahedral Complexes Containing only Monodentate Ligands

(a) In $(Ma_4b_2)^{n\pm}$ type complexes, two isomers are possible : cis – isomers in which two b's have adjacent positions, while in trans isomer two b¢s are opposite to each other. An important example of geometrical isomerism in $(Ma_4b_2)^{n\pm}$ is dichlorotetraammine cobalt (III) ion, $[Co(NH_3)_6Cl]^+$. This complex exists in cis–and *trans* isomers.

In The cis-from, the two Cl^- ions are in (any) two adjacent positions whereas in the trans from the two Cl^- ions are in (any) two opposite positions. *Cis*-isomer has blue-violet colour while *trans*-isomer has green colour.

(b) Complexes of the type (Ma_2b_4) exist in two geometrical forms although the structures of such isomers have not yet been fully established (Fig. 3.39). Typical examples are $[Cc(NH_3)_3(NO_2)_3$ and $[Co(NH_2 - CH_2\ COO)_3]$. For compounds with general formulae Ma_4b_2 and Ma_2b_4 two isomers are possible of cach.

(c) The complexes of the type $[Ma_3b_3]^{n\pm}$ exist in the cis-and tran-isomers. An interesting example of this is trichloro tripyridine rhodium (III), $(Rh\ (Py)_3\ Cl_3]$. Another interesting example is $[Cr\ (NH_3)_3\ Cl_3]$.

This exists, in two forms. In one isomer, the three Cl^- ions are on one triangular face and the three NH_3 molecules are on the opposite triangular face of the regular octahedron. This isomer is known as *facial isomer.* In the other isomer, the Cl^- ions are arranged around on the edges of octahedron whereas the NH_3 molecules around the opposite edges. This isomer is known as *peripheral isomer.*

(d) Mabcdef type complexes : This type of complexes can exist in 15 different isomers each of which would have an optical isomer. An example of such complexes is $[Pt\ (py)\ (NH_3)\ (NO_2)\ (Cl)\ (Br)\ (I)]$. Although, it should exist in 15 isomers, yet its three isomers are actually isolated.

OCTAHEDRAL COMPLEXES CONTAINING MONODENTATE AND SYMMETRICAL BIDENTATE CHELATING LIGANDS

(a) The complexes of the type $[M(AA)_2\ a]^{n\pm}$ (where AA is a symmetrical bidentate chelating ligand and a is monodentate ligand).

Example of this type of complexes are $[Co(en)_2\ Cl_2]^+$, $[Co(en)_2(NO_3)_2]^+$, $[Ir\ (C_2O_4)_2\ Cl_2]^{2-}$ and $[Ir(C_2O_4)_3Cl_2]^{3-}$.

(b) The complexes of $[M(AA)_2\ ab]^{n\pm}$ type (where AA = bidentate ligand and a = monodentate and b = monodentate ligand. These exist in cis- and trans-isomers as shown in Fig. 3.44.

Examples are $[Co(en)_2(NH_3)Cl]^{2+}$ and $[Ru(py)(C_2O_4)NO)]^-$.

(c) The complexes of $[M(AA)\ a_2b_2]^{n\pm}$ type. These exist in cis–and trans–isomers. An important example of these is $[Co(en)\ (NH_3)_2\ (Cl_2)]^+$ which exists is cis- and trans- isomers.

OCTAHEDRAL COMPLEXES CONTAINING UNSYMMETRICAL BIDENTATE CHELATING AGENTS

An important example of these is $[M(AB)_3]^{n\pm}$ in which AB is an unsymmetrical bidentate chelating agent (here A and B represent the two co–ordinating atoms of the ligand).

Octahedral complexes containing optically active bidentate ligands: An important example of these is $[Co(en)\ (pn)\ (NO_2)_2]^+$. Here en and pn denote ethylene diamine and 1, 2– diamino propane respectively.

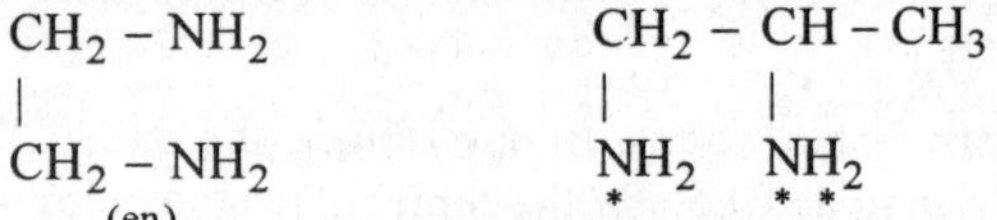

(en) (pn)

As en is symmetrical, it is optically inactive. On the other hand, pn is unsymmetrical and hence it is optically active. In order to differentiate the two donor atoms viz N–atoms of pn; they have shown with *one* and *two* asterisks.

For example, [Co(en) (pn) (NO2)2]+ ion exists in four geometrical forms; two are cis and two are trans. The (a) cis–isomer is obtained by placing $\overset{*}{NH_2}$ and $\overset{**}{NH_2}$ at 4 and 6 positions of the octahedron and the (B) cis–isomer is obtained by reversing their positions, Similarly, (c) and (d) trans isomers.

Polynuclear complexes P The geometric isomers of $[Fe_2(OH)_4(H_2O)^{2+}$. In this example, the two OH^- groups act as bridges and connect the respective octahedral components of the overall structures. The isomerism shown in this example is called bridge isomerism.

To distinguish between Cis and Trans–isomers

The following methods are used to distinguish between cis-and trans- isomers.

Dipole moment method : It is seldom used in coordination chemistry because it does not provide an accurate structural information. However, it is useful in those cases where the presence or absence of dipole moment decides which of two possible structures *i.e., cis* or *trans*-isomers, is actually present.

Jensen showed that the dipole moments are large for cis-isomers and zero for trans-isomers. Jensen studied the platinum complexes of [Pt A_2X_2] type where A = substituted phosphine, arsine or stilbene and X = a halogen and found that their dipole moments (μ) are either equal to zero or between 8 and 12 debye units. In the compounds with (μ =0), the individual components have cancelled one another and so these are trans-isomers. The compounds with μ = 8–12 debye units are *cis*–isomers.

Infra–red technique : As in a *trans*-octahedral complex such as $[Co(NH_3)_4Cl_2]^+$ or in *trans* square planar complex such as $[Pt(NH_3)Cl2]^\circ$ the Cl–metal–Cl symmetrical stretching vibration does not produce any change in the dipole moment of the molecule [Fig. 3.52(b) and (d), no band corresponding to this vibration is observed in the infra-red spectrum. On the other hand, the symmetrical as well as unsymmetrical stretching vibration [Fig. 3.52(a) and (c)] in the *cis*-form of each compound produce appreciable change in the dipole moment. Therefore, the infra-red spectrum of the *cis-isomer* will show a large number of band due to the Cl–metal–Cl stretching.

X-ray structure determinations : X–ray studies have provided proof for the planar structure for many tetra–covalent metals. For example, this study reveals that the anions $[PtCl_4]^{2-}$ and $[PdCl_4]^{2-}$ are planar and that the four nitrogen atoms in the complex cation $[Pt(NH_3)_4]^{2+}$ are coplanar with the Pt(II) atom.

This study also establishes square planar arrangement for 4–coordinated Pt(II), Cu(II), Ag(II) and Au(III). This study also establishes tetrahedral arrangement for Cu(I), Ag(I), Au(I), B(III), Hg(II), Zn(II), Cd(II), Hg(II) and Co(II) in their coordinated compounds. Some metals such as Ni(II) show either configuration for 4-coordination.

As trans isomer possesses plane of symmetry, it cannot be resolved into optical isomers, *i.e.,* it is optically inactive. However, a *cis*-isomer possesses no plane of symmetry and thus can be resolved into d–and l–forms.

Thus the isomer which is optically active is the *cis* isomer and the optical inactive has to be *trans* isomer.

This method is based on the fact that molecular dimensions of bidentate ligands, such, as ethylenediamine, oxalate group and carbonate group are such that they can have only *cis* or 1, 2 positions. Hence, the isomer which is capable of reacting with a chelate group or which is formed by the displacement of such a group must possess cis configuration (provided there is no intramolecular changes during the reaction).

When carbonate tetrammine cobalt (III) chloride $[Co(NH_3)_4CO_3]Cl$, which is a red solid, is treated with cone. HCl, a violet coloured dichloro-tetrammine cobalt (III) chloride $[Co(NH_3)4Cl_2]Cl$, is formed. This compound on keeping for some times slowly changes to green isomer. Since violet compound is obtained from the red solid containing a bidentate ligand (CO_3^{2-}) it has to be the *cis* isomer and the green.

$$\underset{\text{retd}}{[Co(NH_3)_4CO_3]^+} \xrightarrow{HCl} \textit{cis}\text{–}\ \underset{\text{violet}}{[Co(NH_3)_4Cl_2]^+}$$

$$\longrightarrow \textit{trans}\ -\ \underset{\text{green}}{[Co)(NH_3)_4Cl_2}$$

Absorption Spectra : Absorption spectra may also sometimes help in distinguishing *cis* and *trans* isomers of a complex compound. It has often been observed that the absorption spectra of the two isomers differ significantly. A *cis* isomer has a high molar extinction coefficient than the corresponding *trans* isomer.

Kurnakov Reaction : The *trans* effect phenomenon has been used by Russian chemists to distinguish between cis and *trans* isomers of the type $[Pt\ A_2X_2]$. For example, Kurnakov showed that cis $[Pt\ (NH_3)_2Cl_2]$ reacts with thiourea (tu) to yield $[Pt\ (tu)_4]\ Cl_2$ whereas under the same conditions the *trans* isomer would give $[Pt\ (tu)_2Cl_2]$,

$$\text{cis}\ [Pt\ (NH_3)_2Cl_2]+ 4tu \longrightarrow [Pt\ (tu)_4]Cl_2$$

$$\text{trans}\ [Pt(NH_3)_2Cl_2]+ 2tu \longrightarrow [Pt(tu)_2Cl_2].$$

Similar results are reported for reactions of thiosulphate ion with $[Pt (NH_3)_2Cl_2]$; cis isomer forming $[Pt(S_2O_3)_4\text{–}6$ and *trans* isomer yielding $[Pt(NH_3)_2\ (S_2O_3)_2]^{2-}$.

OPTICAL ISOMERISM IN *4*–CO–ORDINATION COMPOUNDS

In terahedral and square complexes of the type Ma_4, Ma_3b and Mab_3, stereo-isomerism is not possible. The reason for this is that all the possible arrangements of the ligands around he central metal ion, M, are exactly equivalent.

Square planar complexes : Optical isomerism is rarely observed in square planar complexes because they have all the four ligands and the central metal ion in the same plane and hence possess a plane or axis of symmetry.

Thus, these complexes cannot show optical isomerism even if all the four ligands are different.

However, Mills and Quibell in 1935 were successful in preparing a four co-ordinate complex of Pt(II) viz iso-butylene diamine-meso-stilbene diamine platinum (II) cation $[Pt(NH_2)CH(C_6H_5)\ CH\ (C_6H_5).NH_2)\ (NH_2\ CH_2\ C\ (CH_3)_2NH_2)]^{2+}$. It has a square-planar shape and has been resolved into two forms. This complex was synthesised as follows:

$$\underset{\substack{\text{iso-butylene}\\ \text{dia min e}}}{\begin{array}{c} H_2C - NH_2 \\ | \\ (CH_3)_2C - NH_2 \end{array}} + (PtCl_4)^{2-} + \underset{\substack{\text{meso-diphenylethylene}\\ \text{dia min e (stillbene)}\\ \text{dia min e)}}}{\begin{array}{c} H_2N - CH(C_6H_5) \\ | \\ H_2N - CH(C_6H_5) \end{array}} \longrightarrow$$

PROOF OF SQUARE PLANAR GEOMETRY

In the above complex cation, there are two rings (I) and (II). If the donor atoms are arranged tetrahedrally around the central Pt^{2+} ion, the ring (II) should be in the plane of paper and the ring (I) perpendicular to it.

But the plane of paper is representing a plane of symmetry. Therefore, the plane of symmetry of the ring (II) will coincide with the plane of ring (I). Thus, this tetrahedral structure possesses a plane or axis of symmetry passing through the plane of ring (II).

Consequently, this complex would not give rise to optical isomers. On the other hand, if the donor atoms are arranged in a square planar around Pt^{2+} ion, this structure will have no plane or axis of symmetry *i.e.* it is asymmetric and consequently optically active. Thus, this complex should give rise to optical isomers.

Octahedral Complexes Containing Monodentate and Symmetrical Bidentate Chelating Ligands

(a) The complexes of the type $[M(AA)_2a_2]^{n\pm}$ (where AA = bidentate chelating agent and a = unidentate ligand) show optical isomerism. An interesting example is dichloro bis (ethylenediamine) rhodium (III). This complex exists in two optical isomers.

Another example of these complexes is $[Co(en)_2Cl_2)^+$ which has two geometrical isomers namely, cis and trans-isomers. The cis-isomer can be resolved into two optical isomers because it does not have plane of symmetry. Thus, trans-isomer is an optically inactive form (mesoform).

(b) The complexes of the type $[M(AA)_2ab]^{\pm}$ (where AA = bidentate ligand and, a and b = mondentate ligands), exist in three forms; two are optically active (d-and l-forms) and the third one is inactive (mesoform).

The examples of this type of complexes are $[Co(en)_2(NH_3)Cl)]^{2+}$ and $[Ru(py)\ (C_2O_4)_2(NO)]^-$

(c) The complexes of the type $[M(AA)a_2b_2]^{n\pm}$ exist in three forms; two are optically active and the third is inactive or symmetrical. Examples are $[Co(en)(NH_3)_2Cl_2]^+$ and $[Co(C_2O_4)(NH_3)_2(NO_2)_2]^-$.

RESOLUTION OF RACEMIC ISOMERS

A racemic isomer is a mixture of optically active d-and l-isomers. *The process of separating a racemic isomer into its d-and l-optically active forms is known as resolution.* Because of the similar physical and chemical properties of the optically active d-and l-isomers, it is not possible to separate them by means of conventional physical methods like fractional distillation of fractional crystallization.

Salt formation method, originally developed by Pasteur in 1858 for organic compounds, is perhaps the best of all methods of resolutions and is widely used for resolving racemic isomer of complex compounds. The method consists in converting the optical isomers in a racemic modification into *diastereoisomers* or *diastereomers,* which are optically active isomers but are not mirror images of each other.

Hence they are not enantiomorphs and therefore possess different solubility in water. These diastereoisomers can then be separated by fractional crystallization.

If the racemic modification is an acid and is treated with an optically active base, two diastereoisomers are formed.

$$(d_{acid} + l_{acid}) + 2d_{\text{base}} \longrightarrow (d_{acid}\ d_{base}) + (l_{\text{acid}}\ d_{base})$$

Racemic form

Similar treatment of basic racemic form with an optically active acid also results in the formation of two diastereoisomers.

$$(d_{base} + l_{base}) + 2d_{acid} \longrightarrow (d_{base}\ d_{acid}) + (l_{base}\ d_{acid})$$

These diastereoisomers have different solubilities, and therefore can be separated by fractional crystallization. After separation the optically active acids or bases used may be regenerated by hydrolysis with inorganic acids or with alkalis. Several resolving agents have been used. Bases which are used for the resolution of racemic acids are mainly alkaloids; *e.g.*, quinine, strychine, brucine, cinchonine, morphine etc. Acids which are used for the resolution of racemic bases are d-antimonyl tartarate ion, d-α-bromocamphor-π-sulphonate anion etc.

The neutral complexes cannot be resolved by this method because they cannot form diastereoisomers. They are therefore resolved by other methods.

The above method of resolution will be explained by taking an example. On treating a solution of a racemic mixture containing d–cis-$[Co(en)_3]Cl_3$ and *l–cis* $[Co(en)_3]Cl_3$ with d–tartarate anion, $(d\text{–tart})^{2-}$ two diastereoisomers, d–cis$[Co(en)_3]Cl_3$ (d–tart) and *l–cis–*$[Co(en)_3]Cl(d\text{–}tart)$ are obtained.

$$[\text{d–cis–}[Co(en)_3]Cl_3 + \text{l–cis–}[[Co(en)_3]Cl_3 + 2(\text{d–tart})^{2-} \longrightarrow$$

Racemic mixture

$$\textit{d–cis–}\ [Co(en)_3]Cl(\text{d–tart}) + \text{l–cis}[[Co(en)_3]Cl(\text{d–tart}) + 4Cl^-$$

The two diastereoisomers have different solubilities in water : the *d–cis* $[Co(en)_3]Cl$ (d–tart) is less soluble in water and hence crystallizes out first in the form of large crystals leaving behind *l–cis* $[Co(en)_3]$ Cl (*d– tart*) in the gelatinous form. These diastereoisomers, on treating with concentrated hydrochloric acid are converted into *d–cis* and *l–cis* isomers and d–tartaric acid is regenerated.

$$\text{d–cis} - [Co(en)_3]Cl\ (\text{d} - \text{tart}) + 2HCl \longrightarrow \text{d–cis} - [Co(en)_3]Cl_3 + \text{d} - \text{tartaric acid}$$

$$\text{l} - \text{cis} - [Co(en)_3]Cl\ (\text{d} - \text{tart}) + 2HCl \longrightarrow \text{l–cis} - [Co(en)_3]Cl_3 + \text{d} - \text{tartaric acid}$$

CO-ORDINATION NUMBER

In a complex compound, the co-ordination number and the geometery (stereochemistry) are dictated by the metal and the ligand. However, it is necessary to point out that the co-ordination number and the resulting stereochemistry are not always predictable. Certain features, however, may be outlined.

1. *As an attraction between a metal ion and dipolar molecules or onions is, involved the coordination number will tend to be as large as possible.*
2. *The arrangement of the co-ordinated ligands must be such that the electrostatic repulsion between ligands is minimied.*

The ML_2 (co-ordination number 2) type complex, therefore should be linear because such a geometry maximises the L–L distance and so minimises the repulsion between the two L groups.

Table 3

Co-ordination Number	*Examples*	*Stereochemistry*
1	$Ag(NH_3)_2^+$ $[CuCl_2]^-$	Linear
2	$[HgI_3]^-$	Distorted equilateral triangle
3	$[CoCl_4]^{2-}$ $[NiCl_2(OPPh_3)_2]$	Tetrahedral
4	$[Pt(NH_3)_4]^{2+}$, $[Ni(DMG)_2]$	Square Planar
5	$[Ru(dipy)_2I]$ I	Trigonal bipyramidal
6	$[Co(NH_3)_6]^{2+}$, $[Cr(en)_3]^{3+}$	Octahedral
	dipy=dipyridyl; en=ethylenediamine	
	$OPPh_3$=trihenylphosphine oxide	
	DMG=dimethylglyoximate ion.	

In ML_3 (co-ordination number 3) complex, the three L ligands are present at the vertices of an equilateral triangle. In ML_4 complex (co-ordination number 4) the four L groups are placed at the corners of a regular tetrahedron or at the corners of a square plane, both stereochemistries assuming minimisation of repulsive forces; ML_5 type complexes have two stereochemistries namely, square-pyramid and a trigonal-bipyramid. In co-ordination number six, this type of complexes admits three stereochemistries, an octahedron, a hexagon or a trigonal prism, but in reality the octahedron is preferred. The Table 3 illustrates stereochemistries with examples :

Complexes with co-ordination Number 2 : Complexes of co-ordination number two are generally formed by copper (I), silver (I), gold (I) and mercury (II). Such complexes possess linear stereochemistry [Fig. 1.47(a)].

Complexes with co-ordination number 3 : The geometrical arrangements, planar and non-planar (trigonal), are possible in the type of complexes with uncommon co-ordination number three. Examples of co-ordination number three are few. The $[HgI_3]^-$ anion is the best studied. In this anion the iodide ions are arranged at the comers of a equilateral triangle with Hg (II).

Complexes with co-ordination number 4 : Please see earlier Co-ordination Number and Stereochemistry.

Complexes with co-ordination number 5: This co-ordination number has been extensively studied now-a-days. A large number of Ni(II) and Co (II) complexes with Schiff s bases (anils) are known $Fe(CO)_5$ $Mn(CO)_5$ have trigonal bipyramidal configurations.

Complexes with co-ordination number 6: Please see earlier Stereo Isomerism.

On the basis of experimental results, the frequency of occurrence of co-ordination numbers for some common ions of some metals are noted below :

Iron(III)	6(Octahedal) > 4(tetrahedral)
Cobalt(II)	6(Octahedral) > 4(tetrahedral) > 5 > 4 (Planar).
Cobalt(III)	6(Octahedral) > > 5, other unknown.
Nickel(II)	6(Octahedral) > (Planar, ~ 4(tetrahedral)
Copper(II)	6(Octahedral) ≡ 5 ~ 4(planar).

FACTORS THAT INFLUENCE COMPLEX FORMATION

It has been observed that the formation of complex ions and coordination compounds is influenced by many physical and chemical factors. These include *environmental factors such as temperature and pressure and more important factors like the nature of the metal ion and nature of the ligand.* The importance of each of these factors can be summarised as follows :

Nature of the Ions outside the Co-ordination Sphere : Sometimes ions like CN^-, SCN^-, CF Br^-, $C_2O_4^{2-}$ and NO_3^- if present outside the complex have a tendency to enter the co-ordination sphere. They can do this at the expense of ligands already co-ordinated to the metal. But ions like NO_3^-, ClO_4^- show little or no tendency to do so.

Role of solvent : Solvent always plays an important role in the synthesis of a particular complex. The selected solvent must be a weaker ligand than the ligand whose complex is to be synthesised. Therefore, syntheses of halo complexes $[NiCl_4]^{2-}$, $[CoCl_4]^{2-}$, etc., are carried out as a rule in non-aqueous solvents like alcohol.

Role of pH : It is a general rule that complexation of a metal ion by a ligand increases with decreasing hydrogen ion concentration, that is, with increasing pH.

Environmental Factors: Temperature and Pressure. The effect of heat can be two fold :

(a) Volatile ligands may be lost at higher temperatures. This is exemplified by the loss of water by hydrates and by the loss of ammonia by $[Co(NH_3)_6]Cl_3$ on heating though this complex ion is inert in strongly acid solution.

$$[Co(NH_3)_6]Cl_3 \xrightarrow{175-180°} [Co(NH_3)_5Cl]Cl_2 + NH_3$$

(b) A second effect of temperature is the transformation of certain co-ordination compounds from one to another. For example, Ag_2HgI_4 is reversibly transformed at 45°C from a red to yellow form.

$$\underset{\text{Red}}{[AgHg[AgI_4]} \rightleftharpoons \underset{\text{Yellow}}{[Ag_2[HgAgI_4]}$$

(c) The effect of pressure is the same as that of concentration. Since the formation of co-ordination compounds involves chemical equilibria, it is apparent that altering the pressure of volatile ligands is merely an application of the Le Chatelier principle.

For example, solid $[Co(NH_3)_6]Cl_2$ loses ammonia upon heating but can be reformed by treatment of anhydrous cobalt (II) chloride with ammonia vapour, that is, by increasing the pressure of ammonia.

$$[Co(NH_3)_6]Cl_2 \xrightarrow{\Delta} CoCl_2 + 6NH_3$$

Effect of Central Ion : It is a special property of d-electrons, that is, of transition metals that they form complexes. The various factors are.

(*a*) *Ionic size* : The smallar an ion, the greater its electrical field, and in general, the more stable its complexes. The following stability constants for hydroxide complexes of the alkaline- earths illustrate the point :

	$BeOH^+$	MgOH+	$CaOH^+$	$BaOH^+$
Stability constant	107	120	30	4

(*b*) *Ionic charge* : For ions of equal size, those with higher charge exert the stronger field and form more stable complexes. The ferricyanide ion ($\log_{10} K = 31.0$), for example, is more stable than the ferrocyanide ($\log_{10} K = 8.3$) and cobalt (II) complexes are more stable than those of the cobalt (III).

Combination of factors (a) and (b) shows that a large charge/radius leads to stable complexes.

(*c*) *Electronegativity* : A strong electron attracting central ion, *i.e.*, one with a high electronegativity, would form the stablest complexes. This is true in some classes only.

Effect of the Ligands : Common ligands possess lone pairs of electrons whereby they serve as donors. Electropositive groups are immediately ruled out as ligands. Ligands containing electronegative atoms such as carbon, nitrogen, oxygen, sulphur or halogens are most useful donors. Some of the characteristics of the ligands include factors such as :

(*a*) *Size and charge of ligand* : For charged ligands, the higher the charge and the smaller the size, the more stable the complex is formed. The F ion, for example, forms more stable complexes than the larger Cl^- ion.

(*b*) *Basic character* : The basicity of a ligand is an important factor because this depends on how easily it can donate electrons. A strong base is a good electron donor, and many common ligands, *e.g.*, CN^-, F^- and NH_3, are also bases. The formation of an acid and a complex have, indeed, something in common

$$H^+ + \text{Base} \rightleftharpoons \text{Acid}$$

$$M^+ + \text{Ligand} \rightleftharpoons \text{Complex.}$$

(*c*) *Extent of chelation* : When metal ions (having co-ordination with molecules or ions which have got more than one point of attachment) form cyclic ring type of complexes called *chelates,* the stability of metal complexes increases. In other words, the formation of a chelate ring increases the stability of a complex over that of a comparable complex where chelation is not possible. For example, $[Cu(en)_2]^{2+}$ is more stable than $[Cu(NH_3)_4]^{2+}$.

(*d*) *Steric effects* : Large, bulky ligands form less stable complexes than similar smallar ones. The ethylene diamine complexes, for example, are more stable than those of the corresponding tetramethyl substituted derivative $(CH_3)_2N.CH_2CH_2N(CH_3)_2$.

(*e*) *Back bonding effect* : The ligands which have back-bonding ability form quite stable complexes. Such ligands are CN, CO, NO_2, C_2H_4, R_3 P, R_3As, R_3Sb and R_2Se. These form stronger complexes with the transition metals.

(*f*) *Orbital effect* : The type and number of orbitals available in the donor atom may also decide the stability of complexes.

COMPLEX FORMATION IN SOLUTION

When a metal ion becomes a part of complex, it achieves new properties which may be strikingly different from those of the original metal ion. In general any property of a system which is related to the concentration of one of the species involved in the formation of the complex may be used to show the formation of a complex. Some other techniques (which have been used) involve such changes which include those in colour, stability towards oxidation or reduction, magnitude of ionic charge (frequently a change in sign) and solubilities and crystalline from of the salts. By careful consideration, one may find one or more suitable methods for the detection of all types of complex formation in the solution.

Ion Exchange Adsorption : This method is especially useful in systems where the sign of ionic charge of the complex is different from that of the metal ion. An anionic charge resin will not adsorb Zn^{2+} from aqueous solution unless a large excess of HCl is present.

In HCl medium, the zinc is in the form of negative ion $[ZnCl_4]^{2-}$ which is strongly adsorbed on the resin. Measurements of the zinc concentrations in the resin and in aqueous phase form a good method for the determination of formation constants of $[ZnCl_4]^{2-}$

Freezing Point Method : It is well known that depression of freezing point is a colligative property as it depends upon the number of the particles present in solution and not on their nature. Thus, this method may be used to detect the formation of complex and to know the number of ions contained in one mole of a given complex compound as the number of particles decreases during complex formation which will cause depression in freezing point.

Change in Chemical Properties : An important method for estabilishing that complexation occurs depends on the ions of the normal chemical properties of the metal ions in solution. The experimental fact that the metal ions do not exhibit their usual reaction is a positive evidence for the complex formation. For example,

(i) *$Fe(OH_2$ does not undergo precipitation upon the addition of OH^- to a solution of* Fe^{2+} *in the presence of EDTA because iron is in the form of complex* $[Fe(EDTA)]^{2-}$.

(ii) *Cr^- does not cause precipitation of Ag^+ from aqueous* NH_3 *solution because Ag^+ ions are present in the form of complex* $[Ag(NH_3)_2]^+$.

Limitations : Usually it is observed that the formation of complexes of low thermodynamic stability cannot be demonstrated by this technique owing to the appreciable amount of simple metal ions left uncomplexed. For example, the complexes between NH_3 and Ca^{2+}, Zn^{2+} or Al^{3+} give all the usual precipitation reactions of the free metal ion.

This method may be used to determine the charge on the complex by measuring the molar conductivity at infinite dilution provided that complex is sufficiently stable.

The knowledge with regard to "the number of ions and the number of charges can be helpful in determining the structure of complex compounds in many cases.

Solubility method : *In some cases, complex formation increases the solubility of the sparingly soluble salt.* For example, when potassium cyanide is added to the solution of sparingly soluble silver cyanide, solubility of silver cyanide increases due to complex formation.

$$\underset{\text{Insoluble}}{AgCN + KCN} \rightleftharpoons \underset{\text{soluble}}{KAg(CN)_2}$$

$$KAg(CN)_2 \rightleftharpoons K^+ + \underset{\text{Complex ion}}{[Ag(CN)_2]^-}$$

Another complex is the formation of $[Ag(NH_3)_2]Cl$ by the action of ammonia on silver chloride

$$\underset{\text{Insoluble}}{AgCl + 2NH_3} \rightleftharpoons \underset{\text{soluble}}{[Ag(NH_3)_2]Cl}$$

If the complex is non-ionic, it is often soluble in a water immiscible organic solvent and the distribution co-efficient can be used as a measure of the amount of complexation.

Dipole Moment : A molecule which is non-symmetrical is said to be a dipole and possesses a permanent electric moment or dipole moment. As applied to co-ordination compounds dipole measurements can be of value in distinguishing between isomers of a compound particularly between *cis* and *irons* isomers. *A irons isomer is having less dipole moment than cis form.*

E.S.R. Spectra : Electron spin resonance spectroscopy is a branch of absorption spectroscopy in which molecules having unpaired electrons undergo transitions between different states when placed in a magnetic field. The absorption takes place in the microwave region and the *energy of* transition $\in$ is given by

$$\in = h\nu = g\beta H$$

where β = magnetic moment in Bohr Magnetons

H = Field Strength

g = gyromagnetic ratio (*i.e.*, the ratio of magnetic moment to angular momentum).

These measurements not only give the value of magnetic moment but also the sub-bands in the absorption bands which have been utilized to explain the metal-ligand bonding (overlap between the orbitals of the metal and ligand) since the unpaired electron placed in a different environment (the electron being delocalised over the ligand) produces a difference in the energy of absorption.

Advantage of E.S.R. method over classical magnetic moment measurement is that it can be carried out with a very small amount of sample and gives much more information and can be used for studying complex formation even if there is no change in the number of electrons of the central metal or ion.

NMR Spectral Measurements : This method is limited to the nuclei–having resultant nuclear spins, transition taking place between different spin energy states of the nucleus by the absorption of frequencies, the nucleus being placed in a strong magnetic field.

These transitions are affected by the presence of unpaired electrons in the orbital of the molecule which may be utilized from measurements of magnetic moments.

In addition, proton n.m.r. and F^{19} n.m.r. have been utilised to study metal ligand bonding in hydrido and fluoro complexes and of ligands having hydrogen, *e.g.,* acetylacetonate (acac).

The shift in the proton n.m.r. absorption (called chemical shift) is a direct proof for complex formation because ring hydrogen absorption is strongly affected by unpaired electron form vanadium (III)

Here also the method can be utilized even if there is no change in the number of unpaired electrons of the central atom or ion on complex formation.

Visible absorption spectral studies : The absorption of light in the visible region (400–800 mμ) by a metal ion depends on the electron transitions within the ion and also on the type and strength of ligand-to-metal bonding.

Thus, the metal ions containing d–electrons (except those having 10 electrons in d-orbitals) give variable colours depending on the nature of the ligands attached with the metal ion.

In many cases the intensity of absorption is increased and the absorption shifts to higher energies (*i.e.*, lower m μ) when coordinated water molecules are replaced by ligands more basic than water.

The visible spectrum shown by dotted lines is that of $[Ni(H_2O)_6]^{2+}$ ion while that shown by solid lines is that of α-amineoxime complex of Ni^{2+} represented as $Ni(AO)_2–H]^+$.

The absorption has shifted to higher energies (*i.e.* lower m μ) and has become more intense.

E.M.F. Method : It has been observed that there is an appreciable change in the E.M.F. of the system whenever complex formation takes place. E.M.F. of any system depends upon the nature of the ion present in it.

Electromigration Studies : Electrophoresis and electrodialysis may be used to detect complex formation in certain cases where the complex ion has a negative charge. Movement of coloured ion (complex metal ion) will be towards the positive electrode instead of negative electrode. Cobalt (III) asparitic acid complex.

Change in Oxidation Potential : The formation of metal chelate or complex is accompanied by a decrease in ionic activity of the metal and hence by an increase in its oxidation potential. The standard oxidation potential or an increase in valence of a number of transition metal ions and their complexes are listed in the Table 4.

In general, the data indicate a consequence increase in the oxidation potential as a result of complex formation. It is apparent that, in general, co-ordination with a donor group increases the oxidation potential and hence increases the relative stability of higher valance state. This is particularly true of combination with negatively charged donor.

Table 4

Reaction	*E° (volts)*
1. (a) $Cu^+ \longrightarrow Cu^{2+} + le^-$	–0.167
(b) $2NH_3 + [Cu(NH_3)_2I_2]^+ \longrightarrow [Cu(NH_3)_4]^{2+} + le^-$	+ 0.580
2. (a)$Co^{2+} \longrightarrow Co^{3+} + le^-$	–1.853
(b) $[Co(NH_3)_6]^{2+} \longrightarrow [Co(NH_3)_6]^{3+} + le$	–0.1
3. (a) $Fe^{2+} \longrightarrow Fe^{3+}\ le^-$	–0.771
(b) $[Fe(CN)_6]^{4-} \longrightarrow [Fe(CN)_6]^{3-} + le^-$	–0.36

Colour Change : sometimes colour change is accompanied by complex formation. For example, when ammonia is added to Cu^{2+} ion, a deep blue colour is obtained due to the formation of $[Cu(NH)_3)_4]^{2+}$, a complex ion.

Amperometry : It involves measurement of currents at constant voltage at the dropping mercury electrode by choosing an electrode potential where only metal ion is reduced. There is decrease in current with an increase in ligand concentration, *i.e.,* decrease in free metal ion concentration and finally attains a constant minimum value.

pH Method : Generally ligands are either weak bases or acids. Thus, the complex formation may be considered to be formed by the displacement of one or more, usually, weak acidic protons of the ligand by the metal ion.

This increases $[H^+]$ ion concentration in complex formation, and, hence, causes a drop in the pH. Thus, the pH effect can be used as an indication of the extent of complex formation.

3

Theory of the Crystal Field and Valence Bond

INTRODUCTION

Colour of Transition Metal Complexes

Most of the complexes of transition metals give absorption bands in the visible region. It means that they are coloured. According to the crystal field theory, the colour is due to the *d-d* transitions between the *eg* and t_{2g} electrons. The energy difference between the e_g and t_2g electrons is so small that even absorption of low radiation in the visible light may cause excitation from a lower to a higher *d* level.

As a result of the absorption of some selected wave lengths of visible light, the complexes appear coloured. For example, the complex $[Ti(H_2O)_6]^+$ absorbs green and yellow light from the white light, it means that the complex will have blue and red colour *i.e.,* purple colour.

The colour intensity of a transition metal complex depends upon the magnitude of Δ. For example, the value of Δ for $[Co(NH_3)_6]^{2+}$ is very large.

It means that it requires relatively large excitation energy which it gets from the blue portions of the visible light. Its aqueous solution, therefore, appears red (white = blue + red).

The magnitude of Δ depends upon electronic configuration, size and oxidation state of metal ion, nature of ligand and geometry of the complex.

Relative Liabilities of Complexes

The crystal field theory gives useful information about relative liabilities of certain complexes and also provides information about the mechanism of their substitution reactions. For example,

(i) From the crystal field stabilisation energy calculations of strong field octahedral complexes, it is evident that the speed of d^0, d^1 and d^2 complexes will be greater than the corresponding d^3, d^4, d^5 and d^6 complexes. This speed is independent of $S_N{}^1$ and $S_N{}^2$ mechanism.

(ii) CFSE calculations reveal a decrease in the order of reactivity of a, d^5, d^4, d^3, d^6 complexes.

(iii) When CFSE calculations are made for weak-field octahedral complexes, it follows front these values that complexes of d^n configuration except d^3 and d^8 should lead to fast reactions irrespective of $S_N{}^1$ and $S_N{}^2$ mechanism.

(iv) CFSE calculations also predicted that d^8 complex (Ni complex) should be inert.

All the above conclusions are in full agreement with the experimental evidences.

Stabilisation of Oxidation States : The crystal field theory explains why certain oxidation states are preferentially stabilised by coordination to specific ligands. For example, water stabilises Co(II) whereas ammonia stabilises cobalt (III).

Magnetic Properties of Complexes: The crystal field theory is successful in calculating the number of unpaired electrons in high spin and low spin complexes. With the help of number of unpaired electrons, one can find the value of 'spin only moment' by applying the formula:

$$\mu_s = \sqrt{[n(n+2)]} \text{ B.M.}$$

From a knowledge of unpaired electrons and spin only moment, it is possible to know two important aspects of complexes.

(i) The valence state of the metal ion in an octahedral complex can be calculated.

(ii) If there are more than 3d electrons, the nature of bonding in a complex may be known.

These two points may be understood by undergoing the following examples.

(i) The magnetic moment of $Hg[Co(CNS)_4]$ is 4.5 B. M. From this value, it is evident that cobalt is in the bivalent state ($Co^{2+} \rightarrow d^7$) and the bonding belongs to spin-free type corresponding to n = 3.

(ii) The magnetic moment of $[Fe\ (diph)_3]\ (ClO_4)_3$ is 2–2 B.M. From this value, it is evident that iron is in the trivalent state ($Fe^{3+} \rightarrow d^5$) and the bonding belongs to spin paired type which corresponds to n = 1.

The values of magnetic moments also help to decide whether a given octahedral complex is a high-spin or a low-spin complex. If the value of μ_{exp} value for the complex is less than the μ, value for the metal ion, the complex belongs to low-spin type otherwise it belongs to high-spin type.

Stereochemistry of Complexes : The crystal field theory explains that copper (II) forms square planar complexes than tetrahedral or octahedral complexes The reason for this is that copper (II) possesses a much higher CFSE. (12.28Dq) in a square planar configuration than octahedral (CFSE = 6Dq) or tetrahedral configuration (CFSE= – 2.56Dq).

The CFSE also explains that the square planar complexes of 4d and 5d metal ions form more readily than *3d* ions.

Table 1

	CFSE (-Dq) Weak field (H_2O)	*CFSE (-Dq) Strong field (NH_3)*
$Co^{2+}(d^7)$	7	18
Co^{3+} (d^6)	4	24

DISTORTED OCTAHEDRAL COMPLEXES

In an octahedral complex, six ligand molecules are arranged around the central metal ion. When the bond lengths are the same for all six ligands, the octahahedral is said to have a regular shape. If the bond lengths are not same in all six ligands, the shape of octahedral complex is distorted. This change in shape is known as *distortion* and the complex is called *distorted octahedral complex.*

In 1937 Jahn–Teller theorem was put forward to explain why certain six coordinated complexes possess distorted octahedral geometry.

According to this theorem,

(i) An octahedral complex is said to have a regular shape if the d orbitals (both t_{2g} and e_g sets) are occupied symmetrically.

(ii) When d-orbitals of the central metal ion of an octahedral complex possess t_{2g} orbitals as unsymmetrical orbitals, there occurs slight distortion from the regular octahedron. This distortion is due to unevenly filling in an octahedral complex because inese orbitals do not point towards the ligands. The slight distortion may occur in such cases when the central metal ion contains 1, 2, 4 or 5 electrons.

(iii) Strong distortion occurs when the e_g orbitals of an octahedral complex are unsymmetrically filled because these orbitals do not point towards the ligands. Due to this strong distortion the octahedral shapes of complexes may change to tetragonal and even to square planar complexes. The high-spin complexes of d^4 and d^9 and low spin complexes of d^7, d^8 and d^9 lead to strong distortion.

The above three postulates (i), (ii) and (iii) which describe the effect of unsymmetrical t_{2g} or e_g orbitals on the shape of an octahedral complex is known as Jahn–Teller effect.

Explanation : The crystal field theory explains Jahn–Teller effect. This can be illustrated by considering the distortion produced by the presence of e_g orbitals in a complex of Cu^{2+} ion (d^9 ion).

This ion possesses the configuration $t_{2g}{}^6\ e_g{}^3$ in both the fields. Thus, there are two possible configurations L :

$$\text{I} \longrightarrow t_{2g}{}^6\ (d_z{}^2)^2\ (dx^2 - y^2)^1$$

$$\text{II} \longrightarrow t_{2g}{}^6\ (d_z{}^2)^2\ (dx^2 - y^2)^1$$

As both the configurations (t_{2g}) orbitals) are completely field, it means that asymmetry is due to the incomplete filling of e_g orbitals. Thus, distortion is mainly due to the repulsion of ligands by the electrons occupying e_g orbitals.

In configuration I the doubly occupied $d_z{}^2$ orbital offers greater shielding to the nuclear charge of the copper than the half filled $dx^2 - y^2$ orbitals. The result of this is that :

(i) The ligands along z-axis which realise lower effective nuclear charge tend to move away from Cu^{2+} nucleus, and

(ii) The ligands along xy plane which experience a higher effective nuclear charge tend to move closer.

From the above two observations (i) and (ii), it is evident that the complex of copper should have four short and two long metal ligand bonds.

If the electronic configuration II is accepted, it is expected that there will be a distortion but in the opposite direction in which the configuration I is predicting. According to conlguration II, the net result is :

(i) The ligands along z-axis tend to move closer to the nucleus and

(ii) The ligands along xy plane would move away.

From the above two results (i) and (ii), it follows that the complex of copper should have four long and two short metal-ligands bonds.

Theoretically, there is no explanation for deciding which of the two possible distortions is more likely. However, experimental results have revealed that distortions are more common in those octahedral complexes in which there are four short and two long bonds. For example,

(i) In the cupric chloride crystal each Cu^{2+} ion is surrounded by six Cl^- ions. Four of them are lying at a distance of 2.30 Å and the other two are at 2.95 Å away.

(ii) In cupric fluoride crystal F^- ions are lying at a distance of 1.93Å and the remaining two F^- ion are at 2.27 Å away.

The more stable octahedral distortion corresponding to configuration I $t_{2g}6$ $(dz)^2$ $(dx^2 - y^2)^1$ is being considered. δ_1 and δ_2 represent the splitting of the e_g^{3-} and t_{2g}^6 – levels respectively. Both δ_1 and δ_2 are much smaller as compared with Δ_0 and also, δ_2 is much smaller than δ_1 *i.e.*,

$$\Delta_0 >>> \delta_1 > \delta_2$$

Under these conditions both t_{2g} and e_g orbitals are split. For t_{2g} electrons there is no net energy exchange because four electrons (d_{yx} and d_{zx}) are stabilised by $4 \times (-1/3\ \delta_2) = -4/3\ \delta_2$ and the remaining two electrons (d_{zy}) get stabilised by $(2 \times 1/3\ \delta_2 + 2/3\ \delta_2) = 4/3\ \delta_2$. Thus,

Net energy for t_{2g} electrons = energy gain + energy loss or Net energy gain for t_{2g} electrons = – 4/3 δ_2 + (4/3 δ_2) = 0.

In the splitting at e_g levels, two electrons are stabilised by $2 \times \left(-\frac{1}{2}\delta_1\right)$ δ_1 while one electron is destabilised by $1 \times \left(+\frac{1}{2}\delta_1\right) = \frac{1}{2}\delta_1$, *i.e.*,

Net energy gain for e_g electrons = energy gain + energy loss

$$= +1/2\ \delta_1 - \delta_1 = -\ \delta_1/2.$$

For the above simple calculations, the net energy change is $\delta_1/2$ which provides the driving force of distortion. This energy is called *John-Teller stabilisation energy.*

Failures of C.F.T.

(i) In CFT model metal-ligand bonding is taken to be purely electrostatic which is quite unrealistic. It provides no account of partial covalent character of the metal- ligand bonds.

(ii) The theory cannot explain for the relative strengths of ligands. For example, it provides no account why OH^- has a smaller splitting than H_2O.

(iii) CFT considers only the metal ion d-orbitals without giving any consideration to other metal orbitals such as s, p_x p_y and p_z and ligand π-orbitals.

(iv) CFT is unable to consider the possibility of the existence of π-bonding in complexes despite of its frequent existence in complexes of metals in low or high oxidation states.

(v) Crystal field theory keeps the total emphasis on the metal orbitals without consideration to the ligand orbitals. Therefore, all properties dependent upon the ligand orbitals and their interaction with metal orbitals are not at all explained.

VBT THEORY

Here we shall discuss only valence bond theory and crystal field theory. This theory is mainly due to Pauling. It deals with the electronic structure of the central metal ion in its ground state, kind of bonding, geometry and magnetic properties of the complexes.

ASSUMPTIONS OF VALENCE BOND THEORY

This theory is based on the following assumptions:

1. The central metal atom or ion (as the case may be) makes available a number of empty s, p and d atomic orbitals equal to its coordination number.

   These vacant orbitals hybridise together to form hybrid orbitals which are the same in number as the atomic orbitals hybridising together. These hybrid orbitals are vacant, equivalent in energy and have definite geometry. Important types of hybridisation occuring in the first row transition metal (3d metals) complexes and the geometry of the complexes.

2. The ligahds have at least one σ-orbital. containing a lone pair of electrons.

3. Vacant hybrid orbitals of the metal atom or ion overlap with the filled (containing lone pair of electrons) o-orbitals of the ligands to form ligand → metal σ-bond (represented as $L \xrightarrow{\sigma} M$). This bond which is generally known as coordinate bond is a special type of covalent bond and shows the characteristics of both the overlapping orbitals. However, it also possesses a considerable amount of polarity because of the mode of its formation.

4. The non-bonding electrons of the metal atom or ion are then rearranged in the metal orbitals (viz. pure d, s or p orbitals as the case may be) which do not participate in forming the hybrid orbitals. The rearrangement of non-bonding electrons takes place according to Hund's rule.

Let us discuss these theories one by one.

Heitler–London Theory

This theory is based upon the pairing and resultant neutralization of opposed electron spins. The fundamental idea about this theory is summarized below.

(i) In order for an atom to enter into chemical combination, it must possess one or more unpaired electrons. *The valency of a species is thus determined by the number of unpaired electrons which it possesses.*

(ii) Although, participation in bond formation by paired electrons is formally ruled out, such electrons may be involved if they can be unpaired without the expenditure of excessive energy.

(iii) A chemical bond results in when the two electrons are paired. For a stable bond formation, it is essential that the two electrons must have opposite spins.

COVALENT BOND NATURE

This is important to remark that the treatment of the covalent bond is far more difficult than the electrovalent bond. In the covalent bond one has to understand what types of forces exist which keep the electron pairs in contact with the two atoms. This idea has been explained on the basis of *Wave Mechanics.* Two wave mechanical theories of the covalent bond have been formulated to explain the nature of the covalent bond. These approaches are.

(A) Valence Bond Method : It is due to the work of *Heitler, London, Slater and Pauling.* In this treatment, a molecule is formed by the overlapping of atomic orbitals and a *molecule is composed of atoms which retain their individual character when linked to the other atoms.*

(B) Molecular Orbital Method : It is due to the work of *Hund* and *Mulliken.* In this treatment, atoms composing of molecular orbitals lose their individual character, *i.e., an electron moves in the field of more than one nucleus. It means that molecular orbitals are polycentric.*

OCTAHEDRAL COMPLEXES

Formation of Molecular Orbitals

In the next step, six atomic orbitals of the central metal ion.

$4s$, $4p_x$, $4p_y$, $4p_z$, $3d_z^2$ and $3d_x^2$ overlap with six groups of composite ligand orbitals formed as in step (b)) forming six sigma bonding and six sigma antibonding molecular orbitals. The bonding molecular orbitals are designated as σ^b – MO's whereas the antibonding molecular orbitals are designated as σ^* – MO's. Thus,

(i) 4_S *and* Σ_a *on overlapping give* σ_s^b *and* σ_s^* *MO's.*

(ii) $4px$ *and* Σ_x *on overlapping give* σ_x^b *and* σ_s^* *MO's.*

(iii) $4py$ *and* Σ_y *on overlapping give* σ_x^b *and* σ_y^* *MO's,*

(iv) *$4p_z$ and Σ_z on overlapping give Σ_y^b and σ_z^* MO's.*

(v) $3dx^2 - y^2$ *and* $\Sigma x^2 - y^2$ *on overlapping give* σ^b $x^2 - y^2$ *and* σ^* $x^2 - y^2$ *MO's.*

(vi) *$3d^2$, and Σ_z^2 on overlapping 'give $\sigma^b z^2$ and $\sigma^* z^2$ molecular orbitals.*

The remaining three orbitals on central metal ion, *i.e.*, $3d_{xy}$, $3d_{yz}$ and $3d_{zx}$ are non-bonding orbitals. Thus, there are is molecular orbitals available for electron filling.

Formation of Group or Composite Ligand Orbitals

It is clear that each of the metal orbitals would be in a position to overlap simultaneously with several ligand orbitals which may in turn belong to different ligands. But the same is not true for ligand orbitals. Before overlapping, the six ligand orbitals combine together linearly to form six group σ-orbitals that would be capable of overlapping with the central metal ion. The determination of such linear combination of ligand o-orbitals is made by inspection method which is a straight forward and commonsense method. The linear combinations of ligand orbitals are called composite ligand orbitals.

(i) If the +ve lobes of a orbitals of all the six ligands in an octahedral complex are pointing towards the central metal ion (Fig. 6.17), then the composite orbital, signified by ($\sigma_x + \sigma_{-x} + \sigma_y + \sigma_{-y} + \sigma_{-z}$) is capable of overlapping with the s orbitals of the metal cation. This composite orbital is designated as Σ_a and its value when normalised is as follows :

$$\Sigma_a = \frac{1}{\sqrt{6}} (\sigma_x + \sigma_{-x} + \sigma_y - \sigma_{-y} + \sigma_z + \sigma_{-z})$$

(ii) Similar to above, for overlapping with p orbitals, the composite orbitals can be find out. For px, it would be $\sigma_x - \sigma_{-x}$. Similarly for p_y and p_z these would be $\sigma_y - \sigma_{-y}$ and $\sigma_z - \sigma_{-z}$. The composite orbitals p_x, p_y and p_z are designated as Σ_x, Σ_y, and Σ_z, respectively. Their values, when normalised, are follows.

$$\Sigma_x = \frac{1}{\sqrt{2}} (\sigma_x - \sigma_{-x})$$

$$\Sigma_y = \frac{1}{\sqrt{2}} (\sigma_y - \sigma_{-y})$$

$$\Sigma_z = \frac{1}{\sqrt{2}} (\sigma_z - \sigma_{-z})$$

(iii) For $dx^2 - y^2$, the composite orbital would be $(\sigma_x - \sigma_{-y} - \sigma_y - \sigma_{-y})$. Its normalised value is

$$\Sigma_x{}^2 - y^2 = \frac{1}{2} (\sigma_x + \sigma_{-x} - \sigma_y - \sigma_{-y})$$

(iv) For $d_z{}^2$, the composite orbitals would be $(2s_z + 2s_{-z} - r_x - s_{-x}\ s_y - s_{-y})$ Its normalised value is

$$\Sigma_z{}^2 = \frac{1}{2\sqrt{3}} (2\sigma_z + 2\sigma_{-z} - \sigma_x - \sigma_{-x} - \sigma_x - \sigma_y - \sigma_{-y})$$

In the above equation the value of 2 before σ_z and σ_{-z} is the value of C $(C\psi_M = C_1\psi_1 + C_2\psi_2)$.

The value of C decides the magnitude of contribution of that atomic orbital towards the formation of molecular orbital.

FILLING OF MOLECULAR ORBITALS

Filling of the molecular orbitals occurs according to the Aufbau's principle. With d^n complexes, the t_{2g} and $e_g{}^*$ orbitals will fill up in just the same way as in crystal field theory. Let us illustrate this filling by considering a d^3 octahedral complex like $[Cr(NH_3J)_6]^{3+}$. In this complex, there is a total of 15 electrons (twelve from six metal ligand orbitals overlapping and three from metal *d* orbitals). These electrons are to be accommodated. Out of 15 electrons, twelve electrons are accommodated in the six bonding MO, *i.e.*, A_{1g} T_{1g} and E_g whereas the remaining three go to the triply degenerate *i.e.*, non-bonding T_{2g} orbitals in the unpaired state.

The only difference is that in ligand field theory, the value of Δ arises due to the different amount of electrostatic repulsion suffered by the various metal *d* electrons from the ligands whereas in molecular orbitals theory, Δ arises due to the metal-ligand covalent bonding which produces one set of non-bonding orbital T_{2g} and another of high energy antibonding orbitals (E_g^*) Thus, in molecular orbital theory, the concept of high and low spin complexes remains same as in crystal field theory.

The magnitude of Δ decides the choice between the high and low-spin complexes. For example, in the hexafluoro complex of Co^{3+} the value of Δ is small and the distribution of electrons between and T_{2g} and E_g^* occurs as $T_{2g,}^4$, E_g^{*2}. This makes it a high spin complex.

This distribution also explains why the Co–F bonds in the complex are not very strong.

The reason for this is that the presence of two electrons in the antibonding orbitals reduces the strength of Co–F bonds.

It is also important to remember that high spin complexes contain electrons in the antibonding orbitals This also explains why such complexes are relatively less stable.

π-Bonding in Octahedral Complexes

The metal orbitals used for π bonding are as follows

$$d_{xy}, d_{yz}, d_{zx} - t_{2g} \text{ or } T_{2g}$$

and

$$p_x, p_y, p_z - t_{1u} \text{ or } T_{1u}$$

From the above it follows that t_{1u} orbitals are involved in o as well as π-overlaps whereas t_{2g} involve in π-overlaps. The ligand orbitals which take part in the π-overlaps are generally *pπ* or *dπ* orbitals. In these the arrow indicates the direction of positive lobes of these orbitals.

The composite ligand orbitals for these can be evolved as before. For example, the composite ligand orbitals of the *px (t1u)* orbitals of the metal would be $\pi_{3x} + \pi_{4x} + \pi_{5x} + \pi_{6x}$. For the d_{xz} t_{2g} orbital of the metal, the composite ligand orbital would be $(\pi_{1z} + \pi_{2z} + \pi_{5z} + \pi_{6x})$. Such types of π-orbitals are present in the ligand like oxides and fluorides. These orbitals have lower energy than the metal-*π* orbitals.

In π-overlaps, eighteen molecular orbitals (six metals orbitals and twelve ligand orbitals, 2π-orbitals on each ligand) are formed. Out of these six are bonding and [T_{2g} (3) and $T_1\pi$ (3) and six are non-bonding [$T_{2g}\pi^*$ (3) and $T_{2g}\pi$ (3)].

It is important to remember that T_{1g}* and T_{1g}* have the same ligand combination as T_{1g} and $T_{2g}\pi$, but with every other sign in the linear combination reversed. Similarly, $T_{1g}\pi^*$ and $T_{1u}\pi^*$ have the same combination as $T_{2g}\pi$ and T_{1u} but with every sign reversed.

The various π-bonding MO formed and the proper metal and ligand orbital combination required for them in an octahedral complex are given in Table 2.

Table 2

MO (Complex)	*Metal orbitals*	*Ligand orbital combinations*
T_{1u} (π)	p_x (t_{1u})	$(\pi_{3x} + \pi_{4x} + \pi_{5x} + \pi_{6x})$
	p_y (t_{1u})	$(\pi_{1y} + \pi_{2y} + \pi_{5y} + \pi_{6y})$
	p_z	$(t_{1u})(\pi_{1z} + \pi_{2z} + \pi_{3z} + \pi_{4z})$
T_{1u} (π)	d_{xy}	$(\pi_{1y} + \pi_{2y} + \pi_{3x} + \pi_{4x})$
	d_{yz}	$(\pi_{3z} + \pi_{1z} + \pi_{5u} + \pi_{6y})$
	d_{zx}	$(\pi_{1z} + \pi_{2z} + \pi_{5z} + \pi_{6x})$

The magnitude of Δ increases according to the nature of the ligands in the following order Strong π donor < weak π–donor < negligible π-interaction < weak π–acceptor < strong π-acceptor.

This confirms the order of spectrochemical series.

Central Metal Ion Orbitals : In an octahedral complex the metal ion is surrounded by six ligands placed along the $+x$, $-x$, $+y$, $-y$, $+z$ and $-$ z axes. We know that there are 9 atomic orbitals on the central metal atom if it belongs to 3*d* series. These are one outer 4s, three outer *p* ($4p_x$, $4p_y$ and $4p_z$) and five inner 3*d* ($3d_x^2 - y^2$, $3d_z^2$ $3d_{xy}$, $3d_{yz}$, $3d_{zx}$) atomic orbitals. These orbitals can be divided into two groups.

(i) The first group includes 4s, $3p_x$, $4p_y$, $4p_z$ $3d_z^2$ and $3d_x - y^2$ atomic orbitals. These orbitals have their lobes lying along the axes. The orbitals by overlapping with ligand orbitals form six metal-ligand σ bonds or molecular orbitals.

(ii) The remaining three atomic orbitals of central metal atom do not participate in σ-bonding and hence remain non-bonding. These can, however, take part in π-bonding in octahedral complexes by overlapping sidewise with π-orbitals of the same six ligands that are approaching the central metal ion from both sides along + x, – x, + y, – y + z and – z axes.

On the basis of the group theory, the metal orbitals are usually classified as

$dx^2 - y^2$, dz^2	e_g	or	Eg
s	a_{1g}	or	A_{1g}
p_x, p_y, and p_z	t_{1u}	or	T_{1u}
d_{xy}, d_{yz} and d_{zx}	t_{2g}	or	T_{2g}

In the above table, it is to be remembered that.

(i) a or A orbitals are always non–degenerate.

(ii) e or E orbitals, are double degenerate.

(iii) t or T orbitals are triple degenerate.

(iv) g expresses the gerade nature of orbitals and

(v) u expresses the ungerade nature of orbitals.

An orbital is called a gerade orbital if on traversing equal distances in a straight line in opposite directions from the centre of symmetry, the sign of wave function does not change.

NOMENCLATURE OF COORDINATION COMPOUNDS

Coordination compounds are named according to the rules suggested by *International Union of Pure and Applied Chemistry*, IUPAC (1976). These rules are given below:

Order of Naming Cations and Anions of an Ionic Complex Compound

(i) If a complex compound iis ionic, *i.e.*, if a complex compound is composed of cations and anions, the name of the cation'is mentioned first and then the name of the anion is written. For Example:

(a) In naming $K_2[PtCl]$, the name of the cation, viz. K^+ is written first and then the name of the anion namely $[PtCl_6]^{2-}$ is mentioned.

(b) In naming $[Co(NH_3)_5Cl]Cl_2$, the name of the cation, $[Co(NH_3)_5Cl]^{2+}$ is written first and then the name of the anion, Cl^- is written.

(c) In $[Pt^{4+}(NH_3)_4Cl_2]^{2+}$ $[Pt^{2+}Cl_4]^{2-}$, the name of the cation, $[Pt^{2+}(NH_3)_4Cl_2]^{2+}$ is writen first and then the name of the anion, $[Pt^{2+}Cl_4]^{2-}$ is mentioned.

(ii) If the complex compound is non-ionic (neutral), the name of the complex compound is written as one word. Thus the name of $[Pt(NH_3)_4Cl_2]^0$ is written as one word.

How to Name the Species Present in Coordination Sphere

In naming the species present in coordination sphere, the ligands are named first and than the name of the central metal atom is mentioned.

How to Name the Ligands?

(i) *How to name neutral ligands?*

(ii) *How to name negative (aniomc) ligands?* The names of negative ligands end in o.

(iii) *How to name positive ligands?* The names of positive ligands which are rare, end in ium.

(iv) *How to name free radicals?* Free radicals are given their own names. These are supposed to be neutral species, *i.e.*, they donot contain any charge on them.

(v) *How to indicate the number of ligands.*

(a) If a complex contains two or more simple ligands of the same type (*e.g.*, Cl^-, Br^-, NO_2^- $C_2O^{2-}{}_4$, NH_3, H_2O etc.), their number is indicated by putting prefixes di (for 2), tri (for 3), tetra (for 4), penta (for 5), hexa (for 6), septa (for 7), oct (for 8), nona (for 9), deca (for 10), undeca (for 11), dodeca (for 12) etc. before their names.

Table 3 : Examples of positive ligands and free radicals

Positive (cationk) ligands : The names of positive ligands end in ium, *e.g.*

NH_4^+	..... Ammonium	O_2^+	..... Oxygenyl
H_3O^+	..... Hydronium	NO_2^+	..... Nitronium
NO^+	..... Nitrosonium (also called nitrosyl, nitrolium or nitroso)	[H3N–NH2]+	..... Hydrazinium
Free radicals. Free radicals are given their own names *e.g.*,			
CH_3	..... Ammonium	O_2^+	..... Oxygenyl
C_6H_5	..... Phenyl	CH_3CH	..... Acetonitrile

(b) If a complex contains two or more complex ligands or organic molecules which already have di, tri etc. in their names, their number is indicated by putting the prefixes bis (for 2), tris (for 3), tetrakis (for 4), pentakis (for 5), hexakis (for 6), hepta (for 7), octa (for 8), nona (for 9), deca (for 10), undeca (for 11), dodeca (for 12) etc. before their names. The name of the complex ligand or organic molecule is written in bracket. Examples of complex ligands are ethylene diamine (en), diethylamine, phosphorus trifluoride, acetylacetonato ion, benzoyl acctonate ion, dipyridyl, trichloro stannyl ion ($SnCl^-_3$), pyridine, methyl amine, triphynyl phosphine etc.

(vi) *How to name bridging ligands :* (a) If a complex has bridging ligands, the prefix μ (mi) is used before the name of each of the ligands. For example

$$[(NH_3)_5Co{-}NH_2{-}Co(NH_3)_4(H_2O)]Cl_5$$

is called pentanunine cobalt (in) u-amido tetrammine aquo cobalt (III) chloride or nonammine aquo (μ-amido) dicobalt (III) chloride.

(b) If a complex has two similar bridging ligands, \li-di is used for its name. For example:

$$\left[\begin{array}{ccc} & OH^- & \\ (H_2O)_4\ \overset{3+}{Fe} & & \overset{3+}{Fe}\ (H_2O)_4 \\ & OH^- & \end{array}\right](SO_4)_2$$

is called tetraquo iron (III) μ-dihydroxo tetraquo iron (III) sulphate or octaquo μ-di-hydroxo if-iron (III) sulphate.

(c) When the same ligand is present as a bridging ligand and as a non-bridging (terminal) ligand, the bridging ligand is named first. In case of the presence of three similar bridging ligands, μ-tri is used before their name. For example

$$\left[\begin{array}{ccc} & CO & \\ (CO)_3\ Fe & CO & Fe\ (CO)_3 \\ & CO & \end{array}\right]$$

is called m-tricarbonyl hexacarbonyl di-iron (0).

(d) If the bridging ligand bridges more than two metal atoms, m_3, m_4 etc. is used before the name of bridging ligand. m_3, m_4 etc. indicates the number of metal atoms between which the bridging ligand makes bridges, *e.g.* since in basic berylium acetate, $Be^{2+}_4 O^{2-}(CH_3COO)_6$, *oxo* ($O^{2-}$) group briges four Be-atoms, in the name of this compound m_4 is used before the name of bridging ligand viz. O^{2-} group. Thus, the name of this compound is m_4-oxo-m-hexa acetato tetra beryllium (II).

(vii) *How to name ambidentate ligands :* See Ambidentate ligands and linkage isomerism discussed on page 18.

(viii) *Order of naming ligands :* If the coordination sphere of a given complex compound contains various types of ligands, these ligands are named in alphabetical order. The prefixes di, tri etc. are not considered while determining this alphabetical order. For example $[Co(NH_3)_4Cl(NO_2)]^+$ ion is named as tetraamine-chloronitrocobalt (III) ion.

The International Union of Pure and Applied Chemistry (IUPAC) publication *Nomenclature of Inorganic Chemistry* (1989), Blackwell scientific Publishers, contains the rules for the systematic naming of coordination compounds. The basic rules are summarized here.

1. The positive ion is named first followed by the negative ion.
2. When writing the name of a complex, the ligands are quoted in alphabetical order, regardless of their charge (followed by the metal).
3. When writing the formula of complexes, ligands are named before the metal. The coordinated groups are listed in the order: negative ligand, neutral ligands, positive ligands (and alphabetically according to the first symbol within each group).

(a) The names of negative ligands end in –o, for example :

F^-	fluoro	H^-	hydrido	HS^-	mercapto
Cl^-	chloro	OH^-	hydrxo	S^{2-}	thio
Br^-	bromo	O^{2-}	oxo	CN^-	cyano
I^-	iodo	O_2^{2-}	peroxo	NO^{2-}	nitro

(b) Neutral groups have no special endings. Examples include NH_3 ammine, H_2O aqua, CO carbonyl and NO nitrosyl. The ligands N_2 and O_2 are called dinitrogen and dioxygen. Organic ligands are usually given their common names, for example phenyl, methyl, ethylenediamine, pyridine, triphenylphosphine.

(c) Positive groups end in-ium, *e.g.* NH_2–NH_2 hydrazinium.

4. Where there are several ligands of the same kind, we normally use the prefixes di, tri, tetra, penta and hexa to show the number

of ligands of that type. An exception occurs when the name of the ligand includes a number, *e.g.* dipyridyl or ethylenediamine. To avoid confusion in such cases, bis, tris and tetrakis are used instead of di, tri and tetra and the name of the ligand is placed in brackets.

5. The oxidation state of the central metal is shown by a Roman numeral in brackets immediately following its name (*i.e.* no space, *e.g.* titanium (III)).
6. Complex positive ions and neutral molecules have no special ending but complex negative ions end in-ate.
7. If the complex contains two or more metal atoms, it is termed polynuclear. The bridging ligands which link the two metal atoms together are indicated by the prefix μ-.

   If there are two or more bridging groups of the same kind, this is indicated by di- μ, tri-μ-etc. Bridging groups are listed alphabetically with the other groups unless the symmetry of the molecule allows a simpler name. If a bridging group bridges more than two metal atoms, it is shown as μ_3, μ_4 , μ_5 or μ_6, to indicate how many atoms it is bonded to.
8. Sometimes a ligand may be attached through different atoms. Thus M–NO_2 is called nitro and M–ONO is called nitrito. Similarly the SCN group may bond M–SCN thiocyanato or M–NCS isothioc–yanato. These may be named systematically thiocyanato–S or thiocyanato–N to indicate which atom is bonded to the metal. This convention may be extended to other cases where the mode of linkage is ambiguous.
9. If any lattice components such as water or solvent of crystallization are present, these follow the name, and are preceded by the number of these groups in Arabic numerals.

STEREOCHEMICAL DESCRIPTORS

Different geometrical arrangements of the atoms attached to the central atom are possible for all coordination numbers greater than one.

The coordination polyhedron (or polygon in planar molecules) may be denoted in the name by an affix called the *polyhedral symbol.* This descriptor clearly distinguishes isomers differing in the geometries of their coordination polyhedra. The same coordination polyhedron, diastereo-isomerism can arise when the ligands are not all alike as with the *cis* and *trans* isomers of tetraamminedichlorochromium (III), diamminedi–chloroplatinum (II), and bis (2–aminoethane–thiolato) nickel (II) (Examples 1–6).

Attempts to produce descriptors similar to *cis* and *trans* for stereochemically more complicated coordination entities have failed to achieve generality, and labels such as *fac* and *mer* are no longer recommended.

Nevertheless, a diastereoisomeric structure may be indicated for any polyhedron using a *configuration index* as an affix to the name or formula. Finally, the chiralities of enantiomeric structures can be indicated using *chirality symbols.*

Polyhedral symbols : The polyhedral symbol indicates the geometrical arrangements of the coordinating atoms about the central atom. It consists of one or more capital italic letters derived from common geometric terms which denote the idealized geometry of the ligands around the coordination center, and an arabic numeral that is the coordination number of the central atom. The polyhedral symbol is used as an affix, enclosed in parentheses, and separated from the name by a hyphen.

Configuration Index

Definition of index and assignment of priority numbers to ligating atoms. Having developed descriptors for the general geometry of coordination compounds, it becomes necessary to identify the individual coordination position.

The *configuration index* is e series of digits identifying the positions of the ligating atoms on the vertices of the coordination polyhedron. The individual configuration index has the property that it distinguishes between diastereoisomers. The digits of the configuration index are established by assigning an order of priority to the legating atoms as described.

The procedure for assigning priority numbers to the ligating atoms of a mononuclear coordination system is based on the standard sequence rule developed for enantiomeric carbon compounds by Cahn, Ingold, and Prelog. The ligating atom with highest priority is assigned the priority number 1, the ligating atom with the next highest priority, 2, and so on.

Configuration Indexes For Particular Geometries

Square planar coordination systems (*SP* – 4). The configuration index in a single digit which is the priority number for the ligating atom *Irons* to the ligating atom of priority number 1.

Table 4 : List of Polyhedral Symbols

Coordination polyhedron	*Coordination number*	*Polyhedral symbol*
linear	2	L–2
angular	2	A–2
trigonal plane	3	TP–3
trigonal pyramid	3	TPY–3
tetrahedron	4	T–4
square plane	4	SP–4
trigonal bipyramid	5	TBPY–5
square pyramid	5	SPY–5
octahedron	6	OC–6
trigonal prism	6	TPR–6
pentagonal bipyramid	7	PBPY–7
octahedron, face monocapped	7	OCF–7
trigonal prism, square face monocapped	7	TPRS–1
cube	8	CU–8
square antiprism	8	SAPR–8
trigonal prism, triangular face bicapped	8	TPRT–8
trigonal prism, square face bicapped	8	TPRS–8
trigonal prism, square face tricapped	9	TPRS–9
heptagonal bipyramid	9	HBPY–9
dodecahedron	8	DD–8
hexagonal bipyramid	8	HBPY–8
octahedron, trans-bicapped	8	OCT–8

Octahedral coordination systems (*OC* – 6). The configuration index consists of two digits. The first digit is the priority number of the ligating atom *trans* to the ligating atom of priority number 1. These two ligating atoms define the reference axis of the octahedron. The second digit of the configuration index is the priority number of the ligating atoms *trans* to the ligating atom with the lowest priority number (most preferred) in the plane that is perpendicular to the reference axis.

FORMULAE AND NAMES FOR CHELATE COMPLEXES DESIGNATION OF THE LIGATING ATOMS IN A POLYDENTATE LIGAND

Donor Atom Symbol as the Index

A polydentate ligand possesses more than one donor site, some or all or which may be involved in coordination. Thus, dithiooxalate anion conceivably may be attached through S or O, and these were distinguished by dithiooxalato-S, S′ and dithiooxalato-O, O′ respectively. If the same element is involved in the different positions, the place in the chain or ring to which the central atom is attached is indicated by numerical superscripts.

The kappa convention. as the complexity of the ligand name increases, a more general system is needed to indicate the points of ligation. In the nomenclature of polydentate chelate complexes, single ligating atom attachments of a polyatomic ligand to a coordination center are indicated by the italic element symbol preceded by a Greek kappa, k. Monodentate ambident ligands provide simple examples, although for these cases the kappa convention is not significantly more useful than the 'donor atom symbol' convention [above].

Nitrogen-bonded NCS [formerly isothiocyanato] is thiocyanato-kN and sulfur bonded NCS is thiocyanato-kS. Nitrogen-bonded nitrite [formerly nitro] is named nitrito-kN and oxygen-bonded nitrite is nitrito-kO, *i.e.*, $[O = N\text{–}O\text{–}Co(NH_3)_5]^{2+}$ is pentaamminenitrito-kO-cobalt (III) ion.

For polydentate ligands, a right superscript numeral is added to the symbol *k* in order to indicate the number of identically bound ligating atoms in the *flexidentate*[2] ligand. Any doubling prefixes, such as bis–, are presumed to operate on the locant index as well. Examples [1 and 2] use tridentate chelation by the linear tetraammine ligand, N,N′–bis(2–aminoethyl)1,2– ethanediamine to illustrate these rules.

Stereo Chemical Descriptors for Chelated Complexes

Stereo-chemical descriptors can be provided for compounds containing chelated ligands but they involve considerations beyond those described above. The polyhedral symbol is determined as in the case of monodentate ligand derivatives. Also, the priority numbers are assigned to ligating atoms as for monodentate ligands. However, a general treatment for the assignment of the configuration index requires the use of priming conventions in order to provide a completely systematic treatment. Thus, for a particular diastereoisomer of $[Co(NH_3)_2(NH_2CH_2CH_2NH_2)Br_2]^+$, the polyhedral symbol is OC–6 and the ligating atom priority number. For this case, there are no additional complications, and the configuration index is assigned in the usual way as OC–6–32.

The classic case of diastereoisomerism that arises among chelate ligand derivatives is the tris(didentate) complexes in which the two donor atoms of the identical ligands are different. Glycinate, $NH_2\ CH_2\ CO_2^-$ and 2–aminoethanethiolate, $NH_2\ CH_2\ CH_2S^-$ illustrate this. For complexes of either ligand, the facial and meridional labels described previously could be applied, but the more systematic configuration indexes are *OC*–6–22 and *OC*–6–21.

Priming convention. The configuration index is especially useful for bis(tridentate) complexes and for more complicated cases. Bis (tridentate) complexes exist in three diastereoisomeric forms which serve to illustrate the utility of a priming convention. These isomers are represented below, along with their site symmetry symbols and configuration indexes. For examples 1, 2 and 3, the two ligands are identical and the ligating-atom priority numbers are indicated.

CHIRALITY SYMBOLS

Symbols R and S

There are two established and well-used systems for chirality symbols and these differ in fundamental ways. The first, the convention for chiral carbon atoms is equally appropriate to metal complexes and is most often used in conjunction with ligand chirality. However, it can be applied to metal centers and has been useful for pseudotetrahedral organometallic complexes when, for example, cyclopentadienyl ligands are treated as if they were monodentate ligands of high priority.

Skew-line Convention for Octahedral Complexes

The second is the skew line convention and applies to octahedral complexes. Tris(didentate)complexes constitute a general family of structures for which a useful unambiguous convention has been developed based on the orientation of skew lines which define a helix. Examples

1 and 2 represent the *delta* and *lambda* forms of a complex, such as $[Co(NH_2CH_2\ CH_2NH_2)^{3+}$.

Chirality symbols based on the priority sequence. The procedure is applied as for tetrahedral, but modified because there is a unique principal axis. The structure is oriented so that the viewer looks down the principal axis, with the ligand having the higher priority closer to the viewer. Using this orientation, the priority sequence of ligating atoms in the [horizontal] plane is examined. If(he sequence proceeds in a clockwise fashion, the chirality symbol *C* is assigned. Conversely, if the sequence is anticlockwise, the symbol *A* is assigned.

POLYNUCLEAR COMPLEXES

Polynuclear inorganic compounds exist in a bewildering array of structural types, such as ionic solids, molecular polymers, extended assemblies of oxoanions both of metals and nonmetals, nonmetal chains and rings, bridged metal complexes, and homo and hetero-nuclear clusters. This section treats primarily the nomenclature of bridged metal complexes and homo-and hetero-nuclear clusters.

Bridging ligands, as far as they can be specified, are indicated by the Greek letter μ appearing before the ligand name and separated by a hyphen. The whole term, *e.g.,* μ-chloro, is separated from the rest of the name by hyphens, as in ammine-μ-chloro-chloro, etc., or by parentheses if more complex ligands are involved. The bridging index, the number of coordination centers connected by a bridging ligand, is indicated by a right subscript, μ_n, where $n > 2$. The bridging index 2 is not normally indicated. Bridging ligands are listed in alphabctic order along with the other ligands, but a bridging ligand is cited before a corresponding nonbridging ligand, as with di–μ–chloro-tetrachloro.

Examples

1. $[CoCu_2Sn(CH)_3)]\mu - (C_2H_3O_2)_2(C_5H_5)]$

   bis(μ-acetato) (cyclopentadienyl)(methyl)–cobaltdicoppertin

2. $[\{Cr(NH_3)_5\}_2(\mu\text{–}OH)]Cl_5$

   μ-hydroxo-bis(pentaamminechromium)(5+)– pentachloride

3. $[[PtCl\{P(C_6H_5)_3\}]_2(\mu - Cl)_2]$ di-μ-chlorobis[chloro (triphenyphosphine)–platinum]

4. $[\{Fe(NO2)2\}2\{\mu\text{–}P(C6H_5)_2)_2]$

bis (μ–diphenylphosphido)bis(dinitrosyliron)

Examples

1. $[Br_4ReReBr_4]^{2-}$ bis(tetrabromorhenate)(Re–Re)(2–)
2. $[Mn_2(CO)_{10}]$ bis(pentacarbonylmanganese)(Mn–Mn) or decacarbonyldimanganese (Mn–Mn).

SPECTROCHEMICAL SERIES

One of the factors which affects the degree of splitting of d–orbitals (10D or Δ) is the *nature of the ligands.* The value of 10Dq can be equated to the energy observed as the maximum in the absorption spectra of complexes as explained earlier for $[Ti(H_2O)_6]^{3+}$. ON this basis, ligands may be arranged in order of their field strength, *i.e.,* the order of energy of the first absorption bands of the complex.

$I^- < Br^- < S^{2-} < SCN^- < Cr^- < NO_3^- < F^- < OH^- < Ox^{2-} <$

$H_2O < NCS^- < NH_3 <$ en < dipy < phen $< NO_2^- < CN^- < CO$

This arrangement of ligands in the order of *increasing field strength* is called Fajans-Tschida *Spectrochemical series* because it is derived from spectral studies of the complexes. The ligands on the right side of the series provide the strongest field (high 10Dq) are the most effective in forcing the cations into low spin states whereas the ligands on the left, provide weak field (low 10Dq) add usually give high spin complexes.

For example we may consider the formation of two octahedral complexes of Fe^{3+} $[Fe(H_2O)_6]^{3+}$ and $[Fe(CN)_6]^{3-}$ where Fe^{3+} has d^5 configuration. Since H_2O is a weaker ligand than CN^- ion, the value of splitting energy (10 Dq) is smaller in $[Fe(H_2O)_6]^{3+}$ than in $[Fe(CN)_6]$. Hence, the five d electrons of Fe^{3+} will occupy all the t_{2g}, and e_g levels without pairing, resulting in a high spin complex with 5 unpaired electrons. Thus $[Fe(CN)_6]^{3-}$ is a highly paramagnetic complex. In case of $[Fe(CN)_6]^{3-}$ the value of 10 Dq is sufficiently high to force pairing of t_{2g} electrons. Thus in $[Fe(CN)_6]^{3-}$ there are two pairs of electrons in two t_{2g} orbitals and single electron in one t_{2g} orbital. Hence it is also paramagnetic in nature but it is less paramagnetic as compared to $[Fe(H_2O)_6]^{3+}$.

The effect of the order of increasing ligand field strength in spectrochemical series, will be clear from the absorption spectra of some complexes of Co^{3+} and Cr^{3+}. We can see that there is a steady, progress in frequency of the absorption maxima as the ligands change with the linking atoms progressing Cl → S → O → N → C corresponding to a progressive increase in the value of 10 Dq. Though, spectrochemical series allows us to rationalise differences in spectra, it creates many serious problems which could not be accounted for, on the basis of crystal field theory. If the ligands arc considered as point charges or dipoles (as assumed by CFT) then anionic ligands should bring out more crystal field splitting than the neutral molecules.

According to CFT, among negative ions F^- is expected to be a stronger ligand than CN^- and among neutral molecules H_2O is expected to be stronger than NH_3 (H_2O has higher dipole moment than NH_3). But in the spectrochemical series, their positions are just the reverse. The position of these ligands can be explained only when π bonding is assumed between the central metal and the ligand (*Molecular orbital theory*).

Since CFT completely neglects π bonding, it cannot account for the position of all the ligands in the spectrochemical series. This failure to interpret the ligand field strength remains as one of the major drawbacks of the crystal field theory.

NATURE OF METAL-LIGAND BONDING IN COMPLEX COMPOUNDS

Wemer was die first to explain the nature of bonding in complex compounds. However, with the advancement of theories of valence, modern theories have been proposed to explain the nature of metal-ligand bonding in complexes. These theories can also explain the colour, geometry and magnetic properties of the complex compounds. This theory was developed by *Heitler* and *London* in 1927. Later, *Pauling* and *Slater* modified this theory to take into account the directional nature of the covalent bond which predicts the accurate geometries of molecules. They further postulated that resonance structures should also be considered.

These modem theories are :

(i) Valence Bond Theory, VBT (due to L. Pauling and J.L. Slater, 1935)

(ii) Crystal Field Theory, CFT (due to H. Brethe, 1929 and Van Vieck, 1932).

(iii) Ligand Field Theory, LFT or Molecular Orbital Theory, MOT (due to J. Van Vieck, 1935).

SIGMA (σ) AND PI (π) BONDS

A covalent bond or molecular orbital is of two types.

(i) Sigma (σ) bond

(ii) Pi (π) bond

When a bond is formed between two atoms by the overlap of orbitals along their axis (end–to–end–overlap), the bond formed is called *sigma* (σ) bond. At the electron clouds overlap along their axis, maximum overlap is possible. Hence, the sigma (σ) bond is a strong bond. Further, electron cloud of a σ bond is symmetrical about the line joining two nuclei concerned. Hence s bond is defined as :

"Bond orbitals which are symmetrical about the line joining the two nuclei are known as s bonds".

These types of bonds are explained below :

(1) Sigma (s) bond : *This type of bond gets formed by the heat on overlapping of half-filled atomic orbitals along the inter-nuclear axes. It is a* **strong bond** *as the extent of overlapping is maximum.*

Types of axial overlap : Various types of axial overlap of s and p electronic orbitals are as follows :

(i) s–s orbitals overlapping

(ii) s–p orbitals overlapping

(iii) s–s orbitals overlapping.

(i) s–s orbitals overlapping or s–s sigma bond : This type of sigma bond is obtained when half-filled s–orbitals of two atoms get overlapped. For example, **the formation of hydrogen molecule** by the overlapping of 1s–orbitals of two hydrogen atoms, has been involving this type of bonding.

(ii) s–p–orbitals overlapping or s–p sigma bond : This type of sigma bond is obtained when half filled s–orbital of one atom gets overlapped with half–filled orbital of other atom. Following examples are given to illustrate this concept :

1. **Formation of Hydrogen fluoride (HF)**

$_9F = 1s^2\ 2s^2\ 2p_x^{\ 2}\ 2p_y^{\ 1}\ 2p_z^{\ 1}$. It is having one half filled $2p_z$ orbital.

$_1H = 1s^1$. It is having one half filled s–orbital.

One lobe of half filled 2pz orbital (dumbell in shape) gets overlapped with s–orbital of H–atom along the internuclear axis. Because of this sp–molecular orbital is produced.

Due to higher electronegativity of F–atom than that of H–atom, the bond pairs of electrons reside more closer to F–atom. Because of this, charge separation occurs.

2. **Formation of water (H_2O molecule) :** O atom ($1s^2\ 2s^2\ 2p_x^{\ 2}\ 2p_y^{\ 1}\ 2p_{z1}$) is having two half filled orbitals while H–atom (1s1) is having one half filled s–orbital.

Half filled 2p orbitals of oxygen undergo overlapping with s-orbitals of each H–atom to form sp sigma bonds.

*The p-orbitals are making an angle of 90° with one another yet the bond angle in water is 105°. It occurs due to t*he repulsive interaction between lone pairs and also between lone pairs and bond pairs.

3. **Formation of ammonia (NH_3) molecule : N** atom ($1s^2\ 2s^2\ 2p_x^{\ 1}\ 2p_y^{\ 1}\ 2p_{z1}$) is having three half filled 2p orbitals. 1H atom ($1s^1$) has one half filled s–orbital.

Half filled 2p-orbitals of nitrogen get overlapped with s–orbital of each H–atom to yield s–p sigma bonds. The p–orbitals are making an angle of 90° with one another. But the bond angle in NH_3 is 108°. It is attributed to the repulsive interactions between lone pair–bond pair–bond pair.

(iii) *p–p orbitals overlapping or pp–p sigma bond :* This type of sigma bond is obtained when half filled p orbital of one atom gets overlapped with half filled p-orbital of other atom. Following example is given for illustration :

Formation of Fluorine Molecule

($_9F$ = 1s2 $2s^2\ 2p_x^{\ 2}\ 2p_y^{\ 2}\ 2p_z^{\ 1}$). One lobe of half filled 2p–orbital of one F–atom is overlapping with one lobe of half filled $2p_z$–orbital of second F–atom along the internuclear axis. Because of this, p-p (σ) molecular orbital is obtained.

Pi (π) bond : *This type of bond is obtained by the sidewise overlapping of half filled atomic orbitals in a direction which is perpendicular to the internuclear axes.* It is a *weak bond* as the extent of overlapping of atomic orbitals has been somewhat small.

Type of sideways or lateral overlap : The only type of side–wave overlapping has been the p–p atomic orbital overlapping which is perpendicular to the inter nuclear axes. Following reveals that the two p-orbitals have been perpendicular to the inter-nuclear axes. On coming closer, these slightly overlap laterally. The upper lobes of the two orbitals tend to overlap above and lower lobes below the inter-nuclear axes. Pi-molecular orbital is having electron cloud above and below the inter-nuclear axis

Formation of oxygen (O2) molecule : O–atom ($1s^2\ 2s^2\ 2p_x^{\ 2}\ 2p_y^{\ 1}\ 2p_z^{\ 1}$) is having two half filled p–orbitals. One half-filled 2p-orbital of one O-atom is overlapping with one half-filled 2p-orbital of another O-atom along inter-nuclear axis to form σ-bond. Second half-filled 2p orbital of one O-atom is overlapping sideways with second half-filled 2p-orbital of another O-atom to yield p-bond. The molecule is a double bond which consists of one sigma and one p-bond. The covalency of oxygen would be two.

Table 5 : Comparison of sigma (σ) and PI (π) bond.

Sigma (σ) bond	*Pi (π) bond*
(i) It is a strong bond as orbitals overlap to greater extent.	(i) It is a weak bond as orbitals overlap to less extent.
(ii) s-orbitals are involved in the formation of a sigma bond.	(ii) s-orbitals are not involved in formation of pi bond.
(iii) One lobe of p-orbital involving in s-bond formation stretches whereas the other lobe reduces in size.	(iii) Both lobes of p-orbitals involving in a p-bond formation remain of the same size.
(iv) This bond is formed by overlap of orbitals along their inter-nuclear axis, *i.e.*, end to end overlapping.	(iv) Pi bond is formed by side to side overlapping of orbitals (Lateral overlapping).
(v) Sigma bond is a strong bond.	(v) Pi bond is a weak bond.
(vi) Electron cloud of a sigma bond is symmetrical about the line joining two nuclei concerned.	(vi) Electron cloud of pi bond is asymmetrical.
(vii) There can be free rotation of atoms around the sigma bond.	(vii) As the electron clouds overlap above and below the plane of the atoms free rotation is not possible around a pi bond.
(viii) It determines the direction and extent of internuclear distance.	(viii) It has no primary effect on the direction of bond but shortens the internuclear distance.

Formation of nitrogen (N_2) molecule : N-atom ($1s^2\ 2s^2\ 2p_x^1\ 2p_y^2\ 2p_z^1$) is having three filled 2p-orbitals. One of these half-filled p-orbitals of one N-atom would overlap with one half-filled 2p orbital of another N-atom along inter-nuclear axis to yield σ bond.

Other two half-filled 2p orbitals of one N-atom would be overlapping sideways with each half-filled 2p-orbital of another N–atom to yield two p-bonds. The molecule has a triple bond consisting of one sigma and two o-bonds. The covalency of nitrogen would be three. Characteristics of sigma (σ) and Pi (π) bonds : The characteristics of sigma and pi bond are given below under therein Table 5.

GEOMETRY AND SHAPES OF MOLECULES

Molecules known to occupy three dimensions are having particulars shapes which are determined by the bonding arrangements of the atoms. The shape of a molecule becomes important in ascertaining its chemical and physical properties, *e.g.*, methane (CH_4) and water (H_2O) are having nearly equal molecular weights but- methane boils at –161° C and water boils at 100° C. Much of this great difference in boiling points has been attributed to the difference in the shapes of these molecules. By the geometrical structure of a molecule, it is meant the three dimensional arrangement of the atoms of groups.

Ionic Compounds are made up of cations and anions which are held together by electrostatic forces of attraction to form a crystal or giant molecule. In a crystal, each cation gets surrounded by a definite number of anions and vice versa. This number has been found to depend upon the size and charge on the ion and decides the geometry of the molecule. An example is that, sodium chloride does not exist as $Na^+\ Cl^-$ but in this crystal, each Na^+ ion gets surrounded by six Cl^- – ions and *vice versa.*

Because of this, sodium chloride is having octahedral geometry. In other words, *ionic compounds in the solid state do not exist as simple molecules but the ions are arranged themselves in a definite three dimensional pattern called lattice.*

On the other hand *Covalent Compounds* generally are made up of simple individual molecules like O_2, CO_2 and HCl. However, in these simple covalent molecules the constituent atoms have been occupying definite positions relative to one another.

The shape of a molecule depending on the arrangement of valencies within the molecule is called the geometry of the molecule. For example, a diatomic molecule, formed by the bonding of two atoms, has a linear shape. In case of other simple molecules, there is always a central atom which forms chemical bonds with other atoms. There are various possibilities in which these bonds may have their mutual *orientations* (or directions) with respect to one another and with respect to central atom.

Thus, depending upon their relative orientations, the molecule, as a whole will have different shapes. Since their shapes may be different they will possess different bond angles between a pair of bonds and bond lengths.

When the bonded atoms are joined by imaginary lines, a space model is obtained. This is termed as the shape of the molecule. Each molecule is having a definite geometry.

REASONS FOR DEFINITE GEOMETRY IN MOLECULES

Atoms react to form molecules so as to attain a state of minimum energy. An atom gets bonded to other atom in a molecule by a pair of electrons between the nuclei of bonded atoms. In polyatomic molecule, there exists more than one electron pair around the central atom. These electron pairs tend to repel each other, as they are similarly charged. As the repulsive interactions gives rise to an increase in energy, the electron pairs will try to remain as far away from one another as possible. Because of this.

(i) All bonded atoms have been occupying such positions (around the central atom) in space where the repulsive forces between them have been minimum.

(ii) The molecule tends to attain minimum energy and maximum stability.

As there can be only one definite orientation of bonded atoms (or electron pairs) around the central atom corresponding to a state of minimum energy, a molecule would be having only one fixed shape or definite geometry.

In order to decide the geometry of the various molecules, various treatments are employed which are as follows :

1. Helferich's Rules
2. VSEPR Theory
3. Hybridisation.

We will now discuss these one by one.

Herferich's Rules : A useful set of empirical rules, called *Helferich's rules,* can be used to arrive at the spatial arrangement of atoms in simple molecules. *In other words, these rules can be used to determine or predict the shapes of simple molecules.* These rules may be summarised as follows :

(*n*) *Molecules or ions of type AX_2 where X may be the same as A or different.* If the central atom A has no unshared electrons, the molecules or ions of the type AX_2 have straight-line structure. For example, carbon dioxide (CO_2) nitrous oxide (N_2O) and oxide (N_3^{-}) ion have liner structures.

$$:\underset{..}{O}:\ :C::\underset{..}{O}:\ :\underset{..}{N}::\overset{+}{N}:\overset{-}{\underset{..}{O}}:\ :\overset{-}{\underset{..}{N}}::\overset{+}{N}:\ :\overset{-}{\underset{..}{N}}:$$

If the central atom A has one or more unshared electron pairs, the molecules or ions of the type AX_2 will have angular shapes.

For example, water (H_2O), hydrogen sulphide (H_2S), and ptannous choride ($SnCl_2$) should be angular, as the following formulae show :

(b) *Molecules or ions of the type AX_3 where X is different or the same as A.* When a does not carry an unshared pair of H electrons, the molecules or ions AX have planar structures with A at the apex. For example, in the BH_3 or BF_3 or CO_3^{-2} OR NO_3^{-} species, the bond angle is 120°. But since, there is no unbonded electron pair, the angle is not decreased but remains 120°.

When A has one unshared pair or electrons, the molecules or ions of the type AX_3 have pyramidal shapes with A at the apex. For example, NH_3, SO_3^{2-} and ClO_3^{-} are pyramidal with bond angle of 109° 28' but the unbonded pairs tend to occupy more space, resulting in decrease in the bond angle between bonded pairs. In NH_3, the N–H bonds lie at 107° to each other.

Molecules or ions of the type AX_4; Molecules or ions of the type AX_4, have tetrahedral structure if the central atom A has no unshared electrons in its other shell and if this shell is of the rare-gas type. If the outer shell is not of the rare-gas type, molecules or ions of the type AX_4 have planar structure.

For example, methane (CH_4) and ammonium ion (NH_4^{+}) have tetrahedral structures whereas the four co-ordinate complexes of Ni (II) and Pd (II) have planar structure.

It should be born in mind that Halferich's rules are merely empirical generalisations that are sometimes useful in arriving at the directed nature of covalent bonds in simple molecules.

Valence Shell Electron-Pair Repulsion Theory (VSEPR Theory): This theory was developed by *Gillespie* and *Nyholm.* According to this theory, the geometry of a molecule depends upon the number of bonding and non-bonding electron-pairs (lone pairs) in the central atom which arrange themselves in such a way that there is a minimum repulsion between them so that the molecule has minimum energy and maximum stability.

Since there can only be one orientation of orbitals corresponding to minimum energy a molecule of a given substance has invariably a definite shape *i.e.* a definite geometry.

Gillespie has proposed the following rules for explaining the shapes of molecules and ions.

Rule 1 : When the central atom in a molecule is surrounded only by bonded electron pairs but not by lone pairs, the molecule will have a regular geometry of shape. The geometry depends on the number of bonded electron pairs as given in the Table 6.

Table 6

Number of electron pairs	*Geometry*	*Bond angle*	*Examples*
2	Linear	180°	BeF_2,$HgCl_2$,CdI_2
3	Trigonal plane	120°	BF_3, BI_3, GaI3
4	Tetrahedral	109° 28¢	CH_4, $SnCl_4$, $SnBr_4$, BF_4^{1-}
5	Trigonal bipyramid	90° and 110°	PCl_5, $SbCl_5$, VO_3^{-1} ,
6	Octahedral	90°	SF_6, $[SiF_6]^{2-}$ $[PbCl_6]^-$
7	Pentagonal bipyramid	72° and 90°	IF_7

Rule 2 : Some molecules may have distorted or irregular geometry if the electron pairs surrounding the central atom are not equivalent, *i.e.* some are bonded while remaining are lone pairs. Lone pair electrons repel adjacent electron pairs more strongly than do bonding electron pairs. Also, the repulsion is expected to increase in the following order :

bond pair — bond pair < bond pair–lone pair < lone pair-lone pair

Rule 3 : The greater the number of lone pairs of electrons on the central atom, the greater, in general will be the tendency of bond angles to contract and hence, greater will be the distortion in the geometry of the molecule. This rule is understood by considering the following series.

$$NO_2^+, 110°; \ddot{N}O_2\ 104°;\ NO_2^-, 115°$$

Rule 4 : The smaller the electronegativity of the central atom, the smaller is the bond angle and hence, smaller is the distortion in the regular shape of molecules. For example,

(i) PI_3, 102° > PBr_3, 101.5° > PCl_3, 100°

(ii) AsI_3, 101.4 > $AsBr_3$, 100.5°>$AsCl_3$, 98.4°

Rule 5 : Bond angles involving multiple bonds are generally larger than those involving only single bonds. However, multiple bonds do not affect the geometry of a molecule.

Rule 6 : Repulsion between electron pairs in filled shell is greater than that in incomplete shell. For example, the bond angle decreases.

$OH_2 \geq SH_2 > SeH_2 > TeH_2$

104.5 92.2 91.0 89.5

Reason : Lone pair of electrons may be defined as that pair which is found in the valence shell of the central atom. It is not involving in bonding (except when required) and gets influenced by the nucleus of the central atom only.

Due to the attractive influence of two nuclei the electron cloud of the shared pair "contracts" and would be occupying a smaller volume. On the other bond the electron cloud of the lone pair is under the attractive influence of only one nucleus "spreads out" and would be occupying a larger volume.

Hence, lone pairs due to their larger electron clouds become more nearer to other lone pairs. This gives rise to maximum repulsion between lone pairs. Due to the small size of the electron cloud of bond pairs, the repulsive force between the two bond pairs would be the least. The repulsion between a lone pair and bond pair would be somewhat in between the two extreme cases.

APPLICATIONS OF VSEPR THEORY

By applying the above mentioned rules of VSEPR theory, one can deduce the regular geometry as well as the distortion in its regular shape in different molecules.

Regular Geometry

According to this theory, a molecule will have a regular geometry if the central atom is surrounded only by orbitals containing shared pairs of electrons (*i.e.*, the bond pairs) and there are no orbitals containing lone pairs (non-bonding) in the valency shell.

Gillespie and Nyholm calculated theoretically (with the help of simple geometry) the regular geometries of molecules containing two to eight pairs of bonding electrons in the valency shell of a central atom.

Two bonding electron-pairs : If there are two bonding pairs of electrons in a molecule, the only way to keep them to the maximum extent is to arrange them at 180° to each other. It yields a linear molecule with a bond angle of 180°. Example is the molecule of beryllium fluoride.

F – Be – F.

Three bonding electron-pairs : If there are three orbitals around the central atom in a molecule, the molecule in this case will be planar triangle and in this case the bond angle should be 120°. This is the case with boron trifluoride.

Here B, the central atom has the electronic configuration : $1s^2\ 2s^2$, $2p^1$ or 2, 3, *i.e.,* it has three electrons in its valency shell. These three valency electrons form three covalent bonds with three fluorine atoms.

F
:F̤̈:　　|
F – B – F

:F̤̈: B :F̤̈:

The molecule has the shape of an equilateral triangle with boron atom lying at the centre and the three atoms lying at the corners of the triangle. This shape is also known as trigonal plane. The bond angle F–B–F is 120°.

Four bonding electron-pairs : If there are four orbitals around the central atom in a molecule, the only way to keep them farthest apart is to have them directed towards the four corners of a regular tetrahedral.

Thus, the molecule will be tetrahedral in shape and the bond angle in this case will be 109.5°.

An example for this is methane. Hence the central atom. C, has the electronic configuration : $1s^2$, $2s^2$, $2p^2$, outer orbit has four electrons in its valency shell . These electrons form four equivalent covalent bonds with 4 hydrogen atoms,

$$\begin{array}{c} H \\ H:\ddot{\underset{..}{C}}:H \\ H \end{array} \quad \text{or} \quad \begin{array}{c} H \\ | \\ H-C-H \\ | \\ H \end{array}$$

As all the four bond pairs are equivalent, the molecule has a regular geometry of *tetrahedron* with a bond angle H–C–H as 109° 28.

Five bonding electron pairs : If there are six bonding electron pairs the only way to keep them farthest apart is to have an arrangement of the type, triogonal bipyramidal. Examples are PCl_5 (g) and PF_5.

Six-bonding electrons–pairs : It there are six bonding electron pairs in a molecule, the molecule will be octahedral in shape and the bond angle will be 90°. An example of this is SF_6. The central atom, 6 flourine atoms. Since all the six bond pairs are equivalent and no valency electron is left unused, the molecule has a regular geometry of an *octahedron.*

Seven bonding electron pairs : If there are seven boding electron pairs, the molecule will be pentagonal bipyramid. An example of this is IF_7. The structure of this molecule can be better understood if we visualise two pyramids placed in such a way that their bases touch one another congruently. The base of the pyramid is a pentagon.

In terms of VSEPR Theory, a pair of electrons in the valence shell, either bonding or lone pair, have been under the influence the structure of the molecule. These electron pairs, are termed as *subtractive* as they are effective in ascertaining the stereo or structural arrangement of atoms in a molecule.

However, the lone pair of electrons due to their more diffuse (spread out) electron clouds bring about larger repulsions. Because of this, bond pairs move away from them and come closer to each other. This causes the decrease in bond angle from the normal expected values. The effect of the lone pairs on molecular geometry in case of the hydrides of the second row non-metals (having lower subtractive electron pairs).

Regular Geometry of Covalent Molecules

A molecule would be having a regular geometry when the repulsive interactions between the electron pairs have been of equal magnitude.

Conditions for regular geometry : A molecule would be having a regular geometry when it satisfies the following conditions.

(i) The central atom of the molecule should be bonded to all similar atoms, *e.g.*, CH_4, CCl_4.

(ii) The central atom should be surrounded by bonded pair of electrons and should not be having lone pair of electrons.

(iii) All the bond lengths between the central atom and the bonded atoms should happen to be the same.

(iv) The central atom should get bonded to other atoms by single covalent bonds (or same type of bonds).

Distorted or Irregular Geometry of Molecules

A molecule would *be* having irregular geometry if the repulsive interactions between the electron pairs (around the central atom) have been not of equal magnitude.

Conditions for irregular geometry : A molecule is said to be having irregular geometry when it satisfies any one of the following conditions :

(i) The central atom of the molecule should not get bonded to all the similar atoms, *e.g.*, $CHCl_3$, CH_3Cl.

(ii) The central atom should possess one or more lone pair of electrons in addition to one or more bond pair of electrons, *e.g.*,

$$\dot{N}H_3,\ H_2\ddot{\underset{..}{O}}$$

(iii) The central atom may get bonded to similar atoms but should possess different bond lengths, *e.g.*, in PF_5 molecule, the equatorial P–F bond length has been 2.04Å whereas the axial P–F bond length has been 2.19Å.

(iv) The central atom should get bonded to other atoms by single as well as multiple bonds, *e.g.*, SO_3.

Prediction of Shapes of Molecules

In order to have an idea about the geometry on the basis of VSEPR Theory, one should proceed as follows :

(1) First of all count the number of valence electrons. Then write down the Lewis structure for the molecule or ion.

(2) From the Lewis structure, find the number of unshared electron pairs and number of bonds which are made by the central atom to the other atoms (The multiple bonds are counted as a single bond).

(3) Find out the number of subtractive electrons pairs by adding the total number of shared pairs and lone pairs around the central atom.

(a) Shapes of Molecules Having Regular Geometry

1. Shape of BeF_2 molecule (AB_2 type)—Linear : In BeF_2, the central atom is $_4Be$ which is having the configuration $1s^2\ 2s^1\ 2p^1$ in the excited state. It is able to form two bonds with two fluorine atoms because of the presence of two unpaired electrons in its valence shell. In the formation of BeF_2 each of these electrons gets shared by two fluorine atoms. Because of this, the beryllium atom gets surrounded by two bonded pair.

Total number of electrons present in the valence shell of Be = 2

Bonds formed by Be = 2

Number of shared pairs of electrons around Be = 2

unshared pair of electrons = 1/2 [2 – 2] = 0

Thus, total number of subtractive electron pairs around central Be atom = 2 + 0 = 2

But the central Be atom gets bonded to similar F–atoms by single covalent bonds and get surrounded by only two shared pair of electrons. Therefore, the force of repulsion among bonded electron pairs would be equal. Hence, the molecule of BeF_2 is having regular linear geometry with bond angle as 180°.

Other examples of molecules having linear geometry include $ZnCl_2$, $MgCl_2$ $BeCl_2$, BeH_2 and $HgCl_2$.

2. Shape of BF_3 molecule (AB_2 type)—Triangular : In BF_3, central atom is $_5B$ which is having the configuration as $1s^2\ 2s^1\ 2p_x^{\ 1}\ 2p_y^{\ 1}$ in the excited state. It is able to make three bonds with three fluorine atoms because of the presence of three unpaired electrons in its valence shell. In the formation of BF_3, each of these electrons gets shared by three fluorine atoms. Because of this, the boron atom gets surrounded by three bond pairs.

$$\overset{\times}{B}{}^{\times}_{\times} + 3\cdot\ddot{\underset{\cdot\cdot}{F}} \longrightarrow :\ddot{\underset{\cdot\cdot}{F}}\ {}^{\times}\overset{\overset{:\ddot{F}:}{\times\cdot}}{B}{}^{\times}_{\cdot}\ \ddot{\underset{\cdot\cdot}{F}}:$$

But the central B atom in BF3 has been bonded to three similar fluorine atoms by single covalent bonds and has been surrounded by only three shared pairs of electrons. Therefore, the force of repulsion among bonded electron pairs would be equal. Thus, the molecule of BF3 is having regular triangular planar geometry. The three B–F bonds get directed towards the corners of an equilateral triangle with each bond angle as 120°.

Other examples of molecules having triangular planar geometry include BCl_3, BH_3 and $AlCl_3$.

3. Shape of CH_4 molecule (AB_4 type)—Tetrahedral : In CH_4 central atom is C which is having the configuration $1s^2\,2s^1\,2p_y^{\ 1}\,2p_z^{\ 1}$ in the excited state. It is able to make four bonds with four hydrogen atoms because of the presence of four unpaired electrons in its valence shell. In the formation of CH_4 each of these electrons gets shared by four hydrogen atoms. Because of this, the carbon atom gets surrounded by four bonded.

$$\times\underset{\times}{\overset{\times}{C}}\times + 4\dot{H} \longrightarrow H_{\times}^{\cdot}\ \underset{\underset{H}{\cdot\times}}{\overset{\overset{H}{\cdot\times}}{C}}{}_{\times}^{\cdot}\ H$$

Total number of electrons present in the valence shell of carbon atom=4.

Number of bonds made by carbon = 4

Number of shared pair of electrons around carbon = 4

Number of unshared pairs of electrons around carbon

$= 1/2[4 - 4] =$ Zero.

Total number of stereoactive electron pairs present around carbon

$= 4 + 0 = 4.$

The central carbon atom in CH_4 molecule gets bonded to four similar hydrogen atoms by single covalent bonds and gets surrounded by only four shared pair of electrons. Therefore, the force of repulsion among bonded electron pair would be equal. Hence, the molecule of CH_4 is having regular tetrahedral geometry. Four C–H bonds get directed towards the four comers of a regular tetrahedron while carbon atom is situated at the c enter of this tetrahedral, each bond angle would be 109° 28′. In CH_4 molecule three C–H bonds are situated in one plane while four the C–H bond is situated behind the plane. Other examples of molecules possessing tetrahedral geometry included CCl_4, $SiCl_4$ and $SnCl_4$.

4. *Shape of SH_6 molecule (AB_6 type)—Octahedral :* In SF^6, central atom is $_{16}S$ which is having the configuration $1s^2\ 2s^2\ 2p^6\ 3s^1\ 3p_x^{\ 1}\ 3p_y^{\ 1}$ 3pz1 $5d^2$ in the excited state. It is able to form six bonds with six fluorine atoms because the presence of six unpaired electrons in its valence shell. In the formation of SF_6, each of these electrons gets shared by six fluorine atoms.

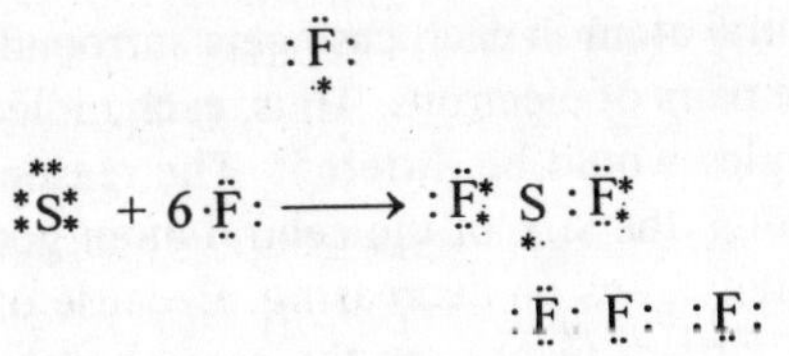

Total number of electrons present in the valence shell of S = 6

Number of bonds made by sulphur = 6

Number of shared pairs of electrons around S = 6

Number of unshared pairs of electrons around S = 1/2 [6 – 6] = 0

Total number of stereoactive pairs present around S = 6 + 0 = 6

But the central sulphur atom in SF_6 gets bonded to six similar fluorine atoms ad gets surrounded by only six shared pairs of electrons. Therefore the force of repulsion among bonded electron pairs would be equal. Thus, the molecule of SF_6 is having regular octahedral geometry. In such a geometry, the central sulphur atom is found in the centre of a square. Out of the remaining two S–F bonds, one is situated above and the other is situated below the square plane and are also perpendicular to it, F–S–F bond angle has been 90°.

Other example of a molecule having octahedral geometry has been TeF_6.

Bond angle in H_2O is less than that in NH_3 : In water molecule two lone pairs of electrons on oxygen are present. In NH_3 molecule, one lone pair of electrons on nitrogen is present. According to VSEPR theory, lone pair-lone pair repulsion > lone pair bond pair repulsion > bond pair-bond pair repulsion. As the number of lone pairs have been more in oxygen, the O–H bond pairs come more closer than N–H bond pairs in NH_3. Hence, bond angle in H_2O (104.5°) would be less than in NH_3 (107°).

Bond angles in the hydrides of elements belonging to the same group : In case of compounds which are differing only in the nature of central atom (the number of electron pairs surrounding central atom and the nature of attached atoms being the same), the bond angles get decreased on moving down the group. For example, the bond angles in the case of hydrides of sixth group elements have been as follows :

Hydride	H_2O	H_2S	H_2Se	H_2Te
Bond angle	104.5°	92.6°	91.0°	90.0°.

The central atom in each case gets surrounded by two bonded pairs and two lone pairs of electrons. Thus, each molecule would be V shaped but bond angles would be different. The reason is that when we move down the group, the size of the central atom goes on increasing and its electronegativity goes on decreasing. Because of this, the bond pairs of electrons lie farther away from the central atom, therefore the force of repulsion between the bond pairs gets decreased. The bond pairs gets decreased. The orbitals having bond pairs will attempt to come closer. Hence, bond angles show the corresponding.

Similarly, it is also possible to explain the decrease in the bond angles in the hydrides of V group elements. The bond angles in these are given below :

Hydride	NH_3	PH_3	AsH_3	SbH_4
Bond angle	107°	93.3°	91.8°	91.3°

Shape of BrF_3 Molecule

In this molecule, there are ten valency electrons in the valency shell of the bromine atom. Out of these, six electrons form three bonding pairs whereas the remaining four electrons form two lone pairs. The expected geometry of BrF_3 is trigonal. But the presence of two lone pairs yields a trigonal bipyramid structure with the lone pairs in equational position. The mutual repulsion of the lone pair binds the axial fluorine atom to give a 'bent T" structure.

Shape of ICl_4^-

In this molecule, there are twelve electrons in the valency shell of iodine. Out of these, eight electrons form four bonding electron pairs whereas the remaining four electrons form two lone pairs. This molecule should have an octahedral geometry with two positions occupied by the lone pairs which can be in the *cis* position in *trans* positions does not give as great angular volume as the *trans* position does. However, the structure of ICl_4^- is square planar with the lone pairs above and below the plane.

Presence of Three Lone Pairs

Formation of ICl_2^- : On the face of it the ICl_2^- anion should have a linear structure. Actually this is found to be so. Although there are three lone pairs, they arrange them-selves at the equatorial positions, causing no distortion. In the centre of this diagram there is iodine atom.

Stability and less reactivity of SF_6 : In SF_6 molecule, all the S–F bonds have been equivalent. The repulsive force between bonded pairs force all atoms to occupy such positions (around the central atom) in space where the repulsive forces between them become equal and minimum. The molecule, therefore, gets minimum energy and maximum stability.

SHAPES OF MOLECULES HAVING IRREGULAR GEOMETRY

Irregular Geometry : Let us now discuss some of the examples in which regular geometries of certain molecules show distortions. Some examples have been described as follows :

1. Shape of PF_5 molecule (AB, type)—Trigonal bipyramidal

In PF_5 central atom is $_{15}P$ which is having the configuration $1s^2\ 2s^2\ 2p^6\ 3s^1\ 3p_x\ 3p_y{}^1\ 3p_z{}^1\ 3d^1$ in the excited state. It is able to make five bonds with the five fluorine atoms because of the presence of five unpaired electrons in its valence shell. In the formation of P F each of these electrons gets shared by five fluorine atoms. Because of this, the phosphorus atom gets surrounded by five bonded.

Total number of electrons present in the valence shell of P = 5

Total number of bonds made by phosphorus = 5

Total number of bond pairs of electrons around phosphorus = 5

Total number of unshared pair of electrons around phosphorus atom

$$= 1/2\ [5 - 5] = 0$$

Total number of stereoactive electron pairs present around phosphorus atom = 5 + 0 = 5

In PF_5, the central phosphorus atom has been bonded to five similar atoms by single covalent bonds and gets surrounded by only five shared pair of electrons. Therefore, PF_5, *should have regular geometry.* However, *actually* the molecule is having *irregular geometry.*

Explanation for irregular geometry : So as to attain minimum force of repulsion among five bonded electron pairs, three FPF bond angles would become 120° and two F–P–F bond angles would become 90°. The molecule gets trigonal bipyramidal geometry. In such a geometry, the central atom (P) is situated in the centre of an equilateral triangle and three P–F bonds (called equatorial bonds) get directed towards the three corners of this triangle. Among the remaining two P-F bonds (called axial bonds), one is situated above and other below the trigonal plane and also perpendicular to it.

Different bond lengths in equatorial and axial bonds : The bond length of equatorial bonds (2.04Å) has been found to be slightly less than axial bonds (2.19Å). *The reason is that, each P–F axial bond faces repulsion from three equatorial bonds at 90° and one axial bond at 180° to it while each equatorial bond faces repulsion from two equatorial bonds at 120° to it and two axial bonds at 90° to it.*

Hence, each axial bond pair (being nearer to three bond pairs at 90°) has to face greater repulsion than PF equatorial bond pairs (which are nearer to two bond pairs at 90°). In order to decrease repulsive force, axial bond pairs tend to move away from the central atom and the bond length of axial P–F bonds gets increased. Because of the difference in bond length values of axial and equitorial bonds, the PF_5 molecule is having irregular geometry.

Unstability and more reactivity of PF_5. The trigonal bipyramidal structure of PF_5 has been an unsymmetrical structure. Because of this, less stable unsymmetrical structure, PF_5 has been highly reactive as compared to SF_6 having symmetrical structure. Other examples of molecules including trigonal bipyramidal geometry include PCl_5, $SbCl_5$, etc.

Presence of One Lone Pair

1. Shape of NH_3 molecule—Pyramidal : In NH_3 the central atom is which is having the configuration $1s^2\ 2s^2\ 2p_x^1\ 2p_z^1$ in the ground state. It is able to make three bonds with three hydrogen atoms because of the presence of three unpaired electrons in it. In the formation of NH_3, each of these electrons is shared by three hydrogen atoms. Because of this, the nitrogen atom is surrounded by three shared pairs of electrons. One lone pair of electrons is still left in the valence shell of nitrogen.

Total number of electrons present in the valence shell of nitrogen=5

Number of bonds made by nitrogen = 3

Number of shared pairs of electrons around nitrogen = 3

Number of unshared pairs of electrons around nitrogen

$$= 1/2(5-3) = 1$$

Total number os stereoactive pairs present around nitrogen

$$= 3 + 1 = 4.$$

The central nitrogen atom in NH_3 gets surrounded by shared as well as unshared as well as unshared pair of electrons. Therefore, the force of repulsion among electron pairs would be unequal. Hence NH_3 molecule would be having *irregular* or *distorted tetrahedral* geometry. It is also called *pyramidal geometry* : In such a geometry, the central atom (N) is situated at the centre of a tetrahedron. The three N–H bonds get directed towards the three corners of this tetrahedron while the fourth corner gets occupied by a lone pair. Each H–N–H bond angle would be 107°.

Explanation for Decrease in Bond Angle

The valence shell of the nitrogen atom in NH_3 contains four electron pairs, as does the carbon atom in methane. Thus, NH_3 might be expected to form a regular tetrahedral configuration. However, the N–H–H bond angle in NH_3 should be 109° 28¢ but actually it is 107°. This decrease in the tetrahedral angle is due to fact that the nitrogen atom contains an unshared electron pair, called a *lone pair.*

Since the lone pair is under the influence of only the nitrogen nucleus, its electron cloud will be spread out in space to a greater extent than will be a bonding electron pair, which is under the influence of both the nitrogen and hydrogen nuclei. This greater repulsion in space of the lone pair electron will result in a greater repulsion between the lone pair and the bond pair than would result between a bond pair and another bond pair. The net result is that the lone pair causes the three N–H bond pairs to be pushed close together, thereby decreasing the bond angle form 109°, 120° to 107°.

Other examples of molecules having pyramidal structure include PH_3,PCl_3 and NF_3. However, the bond angles have been different as compared to the bond angles in case of ammonia.

2. Shape of SF_4 molecule : In SF_4 there are four bonding pairs and one lone pair. If the lone pair is not considered the expected geometry of SF4 should be tetrahedral. But the lone pair has to be considered. In such a situation the geometry should become trigonal bipyrammid . There are two positions for locating the lone pair, *i.e.*, one equatorially. The experimental observed structure shows the position of lone pair equatorially. It causes the repelling of bonding pairs, resulting in the formation of a distorted trigonal bipyramidal structure. In this position the lone pair is under the influence of only two S–F bonds at 90°, whereas in the axial position the lone pair is under the influence of three S–F bonds at 90°. Generally, lone pairs occupy positions where they encounter few 90° interactions.

3. Shape of TeF_5^- **:** In this molecule, there are twelve electrons in the valence shell of Te. Out of these 10 electrons form 5 bonding pairs whereas the remaining two electrons form lone pair. The expected geometry is the octahedral structure whereas the lone pair repulses the fluorine atoms to give a square pyramidal structure.

Presence of Two Lone Pairs

1. Shape of H_2O: bent or V-shaped : In H_2O the central atom is oxygen which is having the configuration $1s^2\ 2s^2\ 2p_x^{\ 2}\ 2p_y^{\ 1}\ 2p_z^{\ 1}$ in the ground state. It is able to make two bonds with two hydrogen atoms because the presence of two unpaired electrons in it. In the formation of H_2O, each of these electrons gets shared by two hydrogen atoms. Because of this the oxygen atom gets surrounded by two shared pairs of electrons. Two lone pairs of electron are still left in the valence shell.

Total number of electrons present in the valence shell of oxygen=6

Number of bonds made by oxygen = 2

Number of shared pair of electrons around oxygen = 2

Number of unshared pair of electrons around oxygen = 1/2 [6 – 2]=2

Total number of stereoactive electron pairs around oxygen in H_2O

= 2 + 2 = 4

In H_2O molecule the central oxygen atom is surrounded by shared as well as unshared pairs of electrons therefore, the force of repulsion among electron pairs becomes unequal. Hence, H_2O molecule is having an irregular or distorted tetrahedral geometry. It is also termed as *bent or angular or V-shaped* geometry. In such a geometry, the central atom (O) is situated at the centre of a tetrahedron. The two O–H bonds get directed towards the two corners of this tetrahedron whereas the two other corners get occupied by two lone pairs. The H–O–H bond angle would be 104.5°.

Explanation for decrease in bond angles : The valence shell of the oxygen atom in water contains four electron pairs and hence it should have tetrahedral structure with bond angle of 109° 28′. But the bond angle in this case is 104° 27′. The reason for this is that two of four electron pairs are lone pairs. From the reasoning used in the case of ammonia, one would expect an even greater repulsion between the electron clouds of the two pairs than the repulsion occurring between a lone pair and a bond pair. This greater repulsion between two lone pair results in a greater pushing together of the bond pair. Hence the H–O–H bond angle in H_2O would be expected to deviate from a tetrahedral angle to a greater extent than the H–N–H angle in NH_3.

Other examples of compounds having bent geometry include H_2S, H_2Se and H_2Te. However, the bond angles have been different as compared to the bond angles in case of water.

Limitations of VSEPR Theory

(i) VSEPR theory does not explain the shapes of molecules having very polar bonds. For example, Li_2O should have the same structure as H_2O but it is linear.

(ii) VSEPR theory does not explain the shapes of molecules which have extensive delocalised π electron systems. For example, $[C(N_3)]^-$ should be pyramidal structure but it is planar.

(iii) VSEPR theory does not explain the shapes of certain compounds which have an inert pair of electrons. For example, AX_6E possessing six bonding and one non-bonding electron pairs should have distorted molecular octahedral geometry but it has regular geometry.

(iv) This theory fails to explain the shapes of certain compounds of transition elements. For example, complexes of transition metals having d^8 configuration are often square planar and not tetrahedral as predicted by Gillespie-Nyholm theory.

(v) VSEPR theory does not make any distinction between the electrons which are present in different types of orbitals, (*i.e.*, *s, p, d* or *f*) of the valence shell.

(vi) VSEPR theory fails to predict the exact bond angles in molecules such H_2O and NH_3 having distorted geometries.

In order to understand this theory completely, one has to understand the following simple examples.

(a) Formation of H_2 molecule : When hydrogen atoms are brought sufficiently close together, it is possible that their 1s charge may interact with each other in one of the two ways.

(i) They can overlap or fuse into each other, resulting in the union of two atoms, or

(ii) They can repel each other and thus prevent any union of the two atoms.

Calculations involving applications of *Pauli's exclusion principle* reveal that the attraction between two atoms occurs when the electrons are of opposite spins, *i.e.*, when they are paired, electrons of the same spins cause repulsion. Thus, for stable bond formation, the condition is that the electrons should have the opposite spins. The concept of orbital overlap can be understood in the following manner. It has been observed that if a mole of H_2 molecule is heated to sufficiently high temperature, the reaction takes place as follows :

$$H_2(g) + 433J \rightarrow H(g) + H(g).$$

It means that, 433 kJ of energy is required to dissociated a mole of hydrogen into hydrogen atoms. In other words, the formation of a mole of hydrogen atoms gives 433 kJ of energy. Thus, the bond formation is accompained by lowering of the energy of the system. In case of the atoms where electron transfer is to favoured on energy consideration, a chemical bond may be formed in some manner which may also lead to lowering of the energy of the system.

For instance, if one or more electrons of one atom may get so close to the other atom that they may experience attractive interaction with the other nucleus (besides its own) the total energy can be lowered. A nucleus at tracts electrons in the region of apace defined as orbital. The overlapping of the orbitals of the two a toms is essential for an electron to face simulataneous attractive interactions from the nuclei of the two atoms. This overlap region should be available to electrons of both the atoms. The occupations of the overlapping orbitals by electrons may take place in the following ways.

(i) Overlapping of orbitals of the atoms which contribute one electron each. The bond formed is called covalent bond.

(ii) Overlapping of orbitals of an atom having two electrons (lone pair) by an empty orbital of an atom. The bond formed is called co-ordinate bond.

Let us now study orbital concept of covalent bonding. The overlapping of s-orbitals in the case of bonding in hydrogen. It can be seen that during overlapping, parts of atomic orbitals of the atoms occupy the same space.

The question now arises, "*how much orbital overlapping can take place?*" It is known that there are attractive and repulsive ineractions present. The repulsive interactions present. The repulsive interactions tend to increase tend to increase the potential energy whereas the attractive interactions tends to decrease the potential energy. Considering the gradual approach of the atoms, at first the attractive interactions increase more rapidly than the repulsive interactions resulting in the decrease of the potential energy of the system.

As the energy of the isolated atom is arbitrarily taken as zero, the decease in the energy of the system is shown as negative. After a certain interatomic distance, the repulsive interactions start increasing more rapidly than the attractive interaction thereby resulting in the increase in potential energy of the system.

The position of minium energy is one where attractive and repulsive forces balance. The more the overlapping of the orbitals, the more will be the decrease in the energy of the system and hence a more stable covalent bond will be formed.

Attractive and repulsive interactions between bonding atoms : The attractive and repulsive interactions between bonding atoms are deciding factors for the formation of a bond between them.

Explanation for the Formation of H_2 Molecule

(i) *Suppose H_A and H_B atoms are at infinite distance from each other.*

In this case, the attractive and repulsive interactions between two atoms have been zero. The potential energy of isolated atoms would be arbi-traily taken as zero.

(ii) *Suppose H_A and H_B atoms are brought closder* : In this case the following attractive and repulsive forces would become operative.

(a) Attractive forces which are existing between electron of H_A and nucleus of H_B and vice versa.

(b) Repulsive forces which are existing between electrons of H_A and H_B and also between the nuclei of H_A and H_B atoms.

But the increase in attractive forces has been greater than the increase in the repulsive forces, it means that energy of the system decrease.

(iii) **Suppose H_A and H_B atoms are at a distance of bond length r_0.**

In this case, three exists net attractive force between atoms. The energy of the system is minium (= d) and a stable H_2 molecule gets formed. Bond length for H_2 molecule = r_0 = 0.74 Å and bond energy = 436.8 kJ/mole.

(iv) **Suppose H_A ad H_B atoms are brought at a distance less than bond length :** In this case, the nuclei of H_A and H_B atoms repel each other strongly. This increase potential energy of the system and curve exhibits upward trend. The molecule is unstable.

He_2 molecule is not formed : There are *two reasons :*

(1) 1s orbital of helium (Z = 2) has been fully filled ($1s^2$). For bond formation, the overlapping orbitals should be half-filled. Thus 1s-orbital of He-atom fails to form He_2-molecule.

(2) (a) *Suppose He_A and He_B atoms are at infinite distance form each other :* In this case the attractive and repulsive interaction between two atoms would be zero. The potential energy of isolated atoms would be thus maximum.

(b) *Suppose He_A and He_B atoms are brought closer :* In this case, following attractive and repulsive interactions come into operation.

(i) *Four new attractive forces* which include two attractive forces between nucleus of He_A–atom and two electrons of He_B–atom (+) and two attractive forces of He_B atoms and two electrons of He_A–atom *i.e.*, 4 attractive forces.

(ii) *Five new repulsive forces :* As the repulsive forces have been more than attractive forces, helium atoms fail to combine to form He_2 molecule.

Formation of Fluorine Molecule : *Fluorine* has electronic configuration as $1s^2\ 2s^2\ 2p^5$. It is difficult to consider its bonding by one orbital similar to that in the hydrogen molecule. The 2p orbitals are split into three sub orbitals under the influence of a magnetic field. These are. These five p electrons are placed, two each in two of p-orbitals and one in the remaining p orbital. The two atoms of fluorine. The two p-orbitals with one electron in each can then form a single bonding orbital with two electrons. It is similar to the hydrogen molecule. In the figure, only p-orbitals are shown.

Formation of Hydrogen Fluoride : Let us consider the hydrogen fluoride molecule, HF. There are five electrons in the 2p orbital of fluorine, *i.e.*, two paired and one unpaired. If the orbital containing the unpaired electrons overlaps the 1s orbital of the hydrogen atom, the spins of the two electrons are opposite. Then just as in the case of hydrogen molecule, the two waves knit together establishing a bond between two atoms.

Successes of the Theory : *The Heitler-London* theory is very useful because many cases of covalent bond formation can be treated in terms of it.

Objectives : Various objections are.

(a) **In the Heitler-London Theory :** it is assumed that the electrons in a shared pair come form different atoms. Thus, the existence of coordinate linkage is ruled out in which electrons come from one atom only and not from two.

(b) This theory fails to explain the structure of perbromates.

(c) The theory fails to explain the structure of odd electron molecules.

(d) In the **Heitler-London Theory :** it is assumed that such paired electrons may be involved in they can be unpaired without the expenditure of excessive energy. In the formation of PF_5, electrons are to be unpaired with required energy. Thus it violates this theory.

(e) It fails to account for the *directional characteristics of covalent jonds*

Reasons for the Failure of Heitler-London Theory : Let us consider two isolated hydrogen atoms. These can be described by the wave functions Ψ_A and Ψ_B, where A and B represent the two hydrogen atoms. If these two atoms are quite far apart, these is no interaction and the wave function for this system of two hydrogen atoms may be given as

$$\Psi = \Psi_A (1)\, \Psi_B (2) \qquad ...(1)$$

where 1 is the electron of atom A and 2 is the electron of atom B.

If we bring the two hydrogen atoms together to form a molecule, they will affect one another and the individual wave function will undergo change. The bonding electrons must be of opposite spin. If they are of parallel spin, no bonding occurs.

By using the wave function, Ψ, as given in equation (1), one can calculate the energy of the system as the two atoms are gradually brought together. Then the energy curve I having a minium of 0.25 eV at an internuclear distance of about 0.9Å is obtained. But the actual experimental bond energy for the hydrogen molecule is 4.7 eV which is much is much higher than this.

Now one may ask questions about the differences between the experimental values and the values obtained from solving equation (1) These differences arise due to the simple fact that equation (1) does not described the hydrogen molecule correctly. A number of limitations have been placed on equation (1). These restrictions were applied by *Pauling and Slater* in 1931. These restrictions are as follows.

(1) First Improvement (Exchange Energy) : As electrons are similar, they are not distinguishable from each other. Hence, it is not proper to lable the electrons as 1 and 2. Even if we are able to distinguish and identify the electrons, we will not be sure that electron 1 was always on atom A whereas electron 2 was always on atom B. Therefore, equation (1) needs modification and it is possible to show the probability that the electrons could exchange their positions so that the electron 1 at any time is associated with atom B and *vice versa.* Mathematically, this situation can be introduced in equation (1) by incorporating a second term *i.e.*,

$$\Psi = \Psi_A (1)\, \Psi_B (2) + \Psi_A (2)\, \Psi_B (1) \qquad ...(2)$$

When equation (2) is solved for the energy associated with it, curve II is obtained which gives bonding energy of 3.14 eV and bond distance of 0.689 Å. The bonding energy, as shown by curve II, compares well with the experimental curve E. The additional lowering in energy has been obtained by virtue of exchange of electrons between the two hydrogen atoms and is known as *exchange energy.* This exchange energy shows that the molecule is more stable and it can also be explained by assuming that now the electron has a larger volume to move.

(ii) Second Improvement (Shielding Effect) : We know that the electrons possess the capacity to shield one another. It means that the effective atomic number Z* is generally less than Z, the atomic number. If one considers the shielding factor, *i.e.*, in which one electron shields the other from the attractive nuclear force, the wave functions Ψ_A and Ψ_B need further changes. When these are considered, the curve III is obtained which approaches the experimental value of curve IV, to a greater extent. The values for the curve III are 3.78 eV for bonding energy and 0.,743 Å for bond distance.

(iii) Third Improvement (Ionicity) : According to equation (2), electrons can exchange places. But exchange takes place simultaneously. It means that if electron 1 is associated with atom B, the other electron has also gone over to A. But this situation is not correct though electron repulsion would tend to keep the atoms apart. Suppose two electrons were present on one atom. It means that the atom would acquire a negative charge and other hydrogen atom would acquire a positive charge.

Thus, there are two ionic structures viz.,

$$H+ - H^- \leftrightarrow H^- H^+$$

These ionic equation have to be added to the s structure as represented by equation (2) which represents a covalent structure, H–H. Thus :

$$H - H \leftrightarrow H^+ H^- \leftrightarrow H^- H^+$$

Therefore, equation (2) becomes as follows.

$$\Psi = \Psi_{A(1)}\ \Psi_{B(2)} + \Psi_{A(2)} + \Psi_{B(1)} + \lambda\ \Psi_{A(1)}\ \Psi_{A(2)} + \lambda\ \Psi_{B(1)}\ \Psi_{B(2)} \quad ...(3)$$

In the above equation, the first two terms represent the covalent structure whereas the remaining two structures represent the ionic structure.

2. **Pauling Slater Theory :** Now we are in a position to summarise the postulates of valence bond theory by taking in to account the above facts. Pauling and Slater developed Heitler-London theory to account for the *directional characteristics in covalent bonds.* To take the geometrical s-apes of orbitals into account and to relate structural chemistry to the covalent bond, the Pauling–Slater's theory suggests that :

(a) The atoms which unite to form a molecule completely retain their identities in the resulting molecule.

(b) A covalent molecule is formed due to the overlapping of atomic orbitals. If there are two atoms each having unpaired electron, the atomic orbitals having unpaired electrons overlap resulting in the formation of a covalent bond between these two atoms in which the spins of two electrons get mutually neutralised. If these atomic orbitals have parallel spins, repulsion will take place and no molecule will, therefore, be formed.

(c) If the atomic orbitals are having more than one unpaired electron, multiple bond formation is possible. In nitrogen molecule, for example, three bonds are possible because N atom has three unpaired electrons.

(d) The number of covalent bonds formed by an atom is equal to the number of unpaired electrons it has.

(e) *The strength of the covalent bond is proportional to the extent of the overlapping of the two atomic orbitals.* The greater the overlap between the atomic orbitals, the greater is the strength of the resulting covalent bond. The extent of overlap of two atomic orbitals of wave function Ψ_A and Ψ_B is represented by overlap integral, S, which is as follows.

$$S = \int \Psi_A \ \Psi_B \ dr$$

From the above relation, it is evident that :

(i) If S is zero, there is no overlap, *i.e.*, there is neither attraction nor repulsion between the combining atoms. Thus, no bond formation is possible.

(ii) If S is positive, there is an increase in the electron density between two nuclei which results in an increase in attraction between the two combining atoms, forming a covalent bond.

(iii) It S is negative, there occurs a decrease in electron density between the nuclei for two combining atoms. Thus, the repulsion between them is increased and hence no bond formation is possible.

(f) *Overlapping takes place between orbitals of only those electrons, which take part is bond formation and not between other orbitals of the atoms.*

(g) *Between two orbitals of the same energy content, the overlapping axially will form a stranger bond.*

(h) *A spherically symmetrical orbital, i.e., s-orbital does not show any directional preference whereas non-spherical orbitals, viz, p, d, f etc., tend to form a bond in the direction of maximum electron density within the orbital, i.e., along their axes.* For example,

(i) As the charge density of s orbital is spherically symmetrical, the two s-orbitals will overlap to the same extent in all directions.

(ii) The three p-orbitals lie on mutually perpendicular axes having nucleus at the centre. Whenever p-orbitals overlap, they tend to form a stronger bond in the direction of their maximum electron density.

(iii) When a s-orbital overlaps p-orbital, the bond is formed but it is weaker than p-p bond but stronger than s-s bond.

Types of Overlapping

Different types of overlapping, which we come across, are :

(a) **s-s overlapping :** This type of overlapping is possible between atoms having half-filled s-orbitals in their outermost energy shells. In the formation of a molecule of hydrogen spherical 1s orbital of one atom overlaps with that of the other forming a single covalent bond.

In the LiH molecule, Li–H bond is formed due to the overlapping of 1s orbital of hydrogen with 2s orbital of lithium.

(b) **s-p overlapping :** This type of overlapping occurs when s-orbital of one atom overlaps with p-orbital of another. For example in the formation of H–F molecule fluorine has atomic number 9 ad its electronic configuration is written as $1s_2\ 2s^2\ 2p^5$. Further p-orbital consists of three sub–orbitals along the three axes, p_x, p_y and p_z,. The complete electronic configuration of fluorine atom is $1s^2\ 2s^2\ 2p_z^2\ 2p_y^2\ 2p_x^2$.

Thus one of the orbital has an unpaired electron available for bond formation. The 1s orbital of hydrogen then overlaps with 2ps orbital of fluorine to form a molecule of HF.

Similarly the molecule of lithium fluoride (Li–F) is produced by the overlapping of 2s orbitals of lithium with 2px orbital of fluorine.

(c) **p-p overlapping :** It is illustrated by the formation of fluorine molecule. The electronic configuration of fluorine is $1s^2$, $2s^2$, $2p_z^2, 2p_y^2\ 2p_x^2$; there is one half-filled 2px orbital. When the overlapping of these two orbitals takes place along their nuclear axis, a sigma bond is formed, as shown below :

Similarly, the formation of Cl–Cl and Br–Br can be explained. Examples. From the concept of valence bond theory, the structure of water (H_2O) and ammonia (NH_3) molecules can be explained.

(a) **Structure of NH_3 molecule :** The electronic configuration of nitrogen is :

$$1s^2\ 2s^2\ 2p_x^1\ 2p_y^1\ 2p_z^1.$$

As the nitrogen atom has three unpaired electrons, it is for the reason that it exhibits a valency of three.

When nitrogen atom combines with three hydrogen atoms to form a molecule NH_3 the three N–H bonds should be at right angles to one another. Thus, the molecule of NH_3 shows a pyramidal structure.

Actually the angle between N–H bonds in NH_3 is 108° instead of 90°. This can be explained from the *polar character* of the bond. As nitrogen has more affinity for electrons than hydrogen, the former (nitrogen) acquires a little negative charge (say, –δ) and the latter (H) a little positive charge (say, + δ). The positively charged hydrogen atoms repel each other and get further apart than 90°.

(b) **Structure of H_2O molecule :** The electronic configuration of exygen atom is given as :

$$1s_2\ 2s^2\ 2p_x^2\ 2p_y^2\ 2p_z^2$$

As oxygen atom has two paired electrons, it means that it has a valency of two. In water molecule (H_2O), the oxygen atom combines with two hydrogen atoms. The two O–H bonds in the formation of water should be at right angles to each other.

Actually the angle between two O–H bonds in H_2O is 105° instead of 90°. Again, it can be explained on the basis of *polar character of the bond.* Due to the great affinity of electrons for oxygen atom, it results in the accumulation of slight positive charge on H atom and a slight –

ve charge on the oxygen molecule repel each other and get further apart than 90°.

An alternative explanation is offered by the concept of *hybridisation.*

Limitations of Valence Bond Method : These are:

(i) It did not explain the formation of molecules such as PCl_5, SF_6 and IF_7 in which the central atom is having 10, 12 and 14 electrons respectively instead of an octet of electrons.

(ii) It did not explain the bonding in electron dificient molecules like B_2H_6; in these molecules, the central atom possesses less electrons than the required octet of electrons.

(iii) It did not explain the existence of paramagnetic molecules like NO, NO_2 and ClO_2,

(iv) It failed to explain the nature of attractive forces that would become operative when a covalent bonds is being formed.

(v) It did not predict the amount of energy which would be involved during the formation of a covalent bond.

(vi) It failed to predict the geometry of covalent molecules like H_2O, CH_4, etc.

(vii) This theory only explains the formation of covalent bond in which the shared pair of electron s comes form two different atoms. However, it offers no explanation for the formation of a co-ordinate bond in which one of the bonded atoms furnishes both the electrons.

(viii) By this method the structure of exygen molecule should be diamagnetic, *i.e.*, all the electrons present on it should be paired electrons. Experimentally, the oxygen molecule has been found to be paramagnetic, indicating the presence of unpaired electrons in oxygen molecule.

(ix) This theory fails to explain the formation of odd electron molecules or ions like H_2^+ ion in which no pairing of electrons occurs.

(x) For any particular molecule or ion, one has to write several canonical structures and even then one is not sure about the true structure which is only a resonance hybrid of these structures. Thus, it is not possible to write one structure which may offer explanation for all the structural features of that molecule or ion.

(xi) It fails to explain the bonding in metals and intermetallic compounds and also in electron deficient molecules.

(xii) It fails to explain the presence of partial ionic nature of a covalent bond.

Inspite of these limitations, the valence bond theory finds wide applications,. It is generally used as a qualitative guide in determining geometry of molecules.

MOLECULAR ORBITAL THEORY

Why molecular orbital theory? In the valence bond method following assumptions are to be made :

(i) Electrons in molecules are localised as if they are in isolated atoms. It means that electron orbital of an isolated atom is not changed if a nucleus of another atom is brought near to it.

(ii) Electrons retain their characteristics even in the presence of other nuclei.

(iii) The wave function of the molecule is the product of the atomic wave functions of the individual atoms.

The above three postulates are not correct and are product of the cases of over simplification. These are taken into account in molecular orbital theory which is therefore more correct.

Valence Bond and Molecular Bond Methods

The valence bond method. mainly due to the work of Heitler, *London, Slater* and *Pauling.* considers the molecule as being made up of atoms with electrons in atomic orbitals on each atom. *Thus, a molecule is treated as if it were composed of atoms which, to some extent, retain their individual character when linked to other atoms.*

The molecular orbital method– mainly due to the work of *Hund, Jones* and *Mulliken– treats a molecule in the same way as an atom, except that in the molecule an electron moves in the field of more than one nucleus, i.e., molecular orbitals are polycentric.*

A and are two isolated atoms. In A and B combine in accordance with valence bond method, *i.e.*, atoms A and B do not lose their individual character while forming a molecule AB. In atoms A and B combine to form molecule AB according to the molecular orbital method.

In this case two electrons move in a field of two nuclei, *i.e.*, molecular orbital are polycentric. Thus, in molecular orbital method, atoms A and B lose their individual character.

The *basic principles of molecular orbital theory* are given as follows:

1. When nuclei of two atoms come close to each other, *their atomic orbitals interact resulting in the formation of molecular orbitals.* The atomic orbitals of the atoms in a molecule completely lose their identity after the formation of molecular orbitals.
2. Each molecular orbital may be described by a wave function Ψ, known as molecular orbital wave function.
3. Ψ is such that Ψ^2 represents the probability density of electron density.
4. Each molecular orbital wave function (Ψ) is associated with a set of quantum numbers which ascertain the energy and the shape of the molecular orbital.
5. Each Ψ is associated with a definite energy value and the total energy of the molecules is the sum of the energies of the occupied molecular orbital.
6. Electrons tend to fill the molecular orbitals in the same way as they fill the atomic orbitals following the Aufbau principle, Pauli's exclusion principle and the Hund's rule of maximum multiplicity.
7. Each electron in a molecular orbital belongs to all the nuclei present in the molecule.
8. Each electron moving in a molecular orbital is having a spin of $+\frac{1}{2}$ or $-\frac{1}{2}$.

The basic difference *between an atomic orbital and a molecular orbital* is that an electron in an atomic orbital is influenced by *one positive nucleus only* while an electron in molecular orbital is influenced by two or more nuclei depending upon the number of atoms contained in a molecule.

There are two methods for getting an approximate yet at the same time a satisfactory idea about molecular orbitals.

1. Linear combination of atomic orbital method (LCAO) method.

2. United atom method.

Let us discuss these one by one.

1. *LCAO Molecular Orbital Theory* : The molecular orbitals can be obtained by the method of linear combination of atomic orbitals (LCAO).

In LCAO method two nuclei must approach one another along a line. When they come close to one another, two atomic orbitals (one form each bonded atom) having comparable energies and possessing large overlap combine to form two molecular orbitals.

One of the new molecular orbitals has lower energy than the combining atomic orbitals and the other has higher energy than the combining atomic orbitals.

The molecular orbital with higher energy than the atomic orbitals gives rise to a repulsive state and is called *antibonding orbital,* while the molecular orbital with lower energy than atomic orbitals gives rise to an attractive state and is called the *bonding orbital.*

It may be further pointed out, that electrons placed in the antibonding orbital tend to increase the energy of molecule, thus decrease its stability and the electrons placed in bonding orbitals tend to decrease the energy of the system, and thus increase its stability.

The bonding and antibonding orbitals are distinguished from each other by putting an asterisk (*) on the antibonding orbitals.

Let us illustrate the formation of molecular orbitals in hydrogen molecule. Before we discuss this theory, we will attempt to explain the meaning of an electron wave (Ψ).

The electron wave above the imaginary horizontal line is given a + ve sign while that below a line is –ve. One should not confuse these sign with the signs of charged bodies. These simply depend upon the value of co-ordinates.

In the formation of hydrogen molecule, s orbital of one hydrogen atom overlaps 1s orbital of another hydrogen atom to form two molecular orbitals. Recalling the wave nature of electrons there are two ways of combining atomic orbitals, (*i.e.*, electron waves).

In this case both the electron waves are on the same side (+ ve side) and on adding result in the formation of a molecular orbital of the type as shown on the right hand side. In such a molecular orbital, the region between the two positive nuclei where 1s orbitals overlap will have high probability.

As the electron density between the two nuclei is maximum, it shields the nuclei from mutual repulsion and holds them at a certain distance from each other which is actually equal to the bond length. Such a bonding orbital is called a bonding molecular orbital and by convention it is designated as σ (1s) orbital. The sign σ signifies that the orbital is symmetrical about the molecular axis joining the two nuclei and the script 1s indicates that molecular orbital is formed by the combination of two 1s atomic orbitals.

Here the two electron waves are oriented in two opposite directions. Therefore, they cancel out each other and thus the probability of finding electrons in the region of overlap is practically nil. As there are no electrons in this region, there is not shielding from nuclear repulsion. Hence there is no bonding and the nuclei are repelled from each other.

Such a molecular orbital is known as antibonding molecular orbital and is designated as σ* (1s). the sign * (asterisk) signifies the antibonding nature of molecular orbital.

A bonding molecular orbital (BMO) : *on account of minimising of repulsion between the nuclei, is associated with much lower energy than either of the atomic orbitals from which it is derived.* An anti-bonding molecular orbital (ABMO), *on the other hand, on account of greater repulsion between the nuclei, is associated with higher energy than the uncombined AOs.*

(a) Features of Bonding and Antibonding Molecular Orbitals : Suppose Ψ_A and Ψ_B are the wave functions of two atoms A and B respectively. If the lobes of the combining atomic orbitals with the same sign overlap, the bonding molecular orbital Ψ_b, is formed..

$$\Psi_b = \Psi_A + \Psi_B \qquad ...(1)$$

When the lobes of the combining atomic orbitals of opposite signs overlap the new molecular orbital is obtained. It is denoted by Ψ_a, the antibonding orbital.

$$\Psi_a = \Psi_A - \Psi_B \qquad ...(2)$$

The electron distribution in a molecular orbital can be obtained by squaring the wave functions. On squaring equations (1) and (2), we get for bonding and antibonding orbitals such as

$$\Psi_b^2 = \Psi_A^2 + \Psi_B^2 + 2\Psi_A \Psi_B \qquad ...(3)$$

and $$\Psi_a^2 = \Psi_A^2 + \Psi_B^2 - 2\Psi_A \Psi_B \qquad ...(4)$$

The equation (3) and (4) represent the probability functions for bonding and antibonding molecular orbitals. These two differ by the cross term $2\Psi_A \Psi_B$. This integral $\int \Psi_A \Psi_B$ is the known as the overlap integral, S. The value of S is numerically very small and, therefore, is neglected in calculating the wave functions of bonding and antibonding orbitals. When the overlap integral S is completely neglected, the molecular wave functions become as follows :

(i) For bonding orbitals,

$$\Psi_b = \frac{1}{\sqrt{(2)}} (\Psi_A + \Psi_B)$$

(ii) For antibonding orbitals,

$$Y_a = \frac{1}{\sqrt{(2)}} (Y_A - Y_B).$$

(b) One can make the clear distinction between bonding and antibonding molecular orbitals by showing the *three-dimensional contours* of electron charge densities or probability densities (viz. Ψ^2) around the two nuclei A and B for these orbitals (a) and (b).

In a bonding MO, the high electron charge density contours are localised on the two nuclei but tend to polarise *towards the space* located in between the two nuclei. The low electron charge density contours are, however, spread over both the nuclei. This implies enhancement of electronic charge in space in between the two nuclei.

The electron density in the bonding orbital tends to move in the region around the two nuclei. Due to the high electron density, the nuclei are shielded from each other and there is greater attraction of the nuclei for the electrons. This results in a lowering of energy and is therefore a bonding structure.

In an antibonding MO, the high as well as low electron charge density contours are localised on the two nuclei and tend to polarise *away from the space* falling in between the two nuclei. This implies depletion of electronic charge from the space in between the two nuclei.

In an antibonding orbital the electrons are in those regions in space where they are not under the attraction of both the nuclei. *The probability of finding the electron in between the nuclei is negligible. The electron density is concentrated on any one nucleus at a time. Every electron in an antibonding molecular orbital contributes to repulsion of the two atoms because it possesses higher energy than the atomic orbitals from which it is formed.*

ENERGY LEVELS OF MOLECULAR ORBITALS

In the method of molecular orbitals, the *s, p* and *d* orbitals in a simple atom are replaced by σ, π and δ orbitals in a molecule. Let us consider the combination of various orbitals to form molecular orbitals :

(i) **Combination of s-orbitals :** A combination of two similar 1s–atomic orbitals will yield two molecular orbitals to accommodate the available electrons. One of the molecular orbitals is a bonding orbital (σ 1 *s*) and the other is antibonding (σ* 1 *s*).

Combination of two 2s atomic orbitals will produce a *2s* and a* *2s* molecular orbitals. These will be of higher energy than o Is and O* Is orbitals since they (σ 2s and σ* 2s) are formed from atomic orbitals of higher energy.

Combination of p-Orbitals : p-orbitals can overlap, and be combined either end to end, to give σ-molecular orbitals or broad side on to give π-molecular orbitals. Let us consider the various p-orbitals.

(ii) *Combination of* $2p_x$ *Orbitals :* Two $2p_z$, atomic orbitals overlapping end to end can form a bonding molecular orbital as well as antibonding molecular orbital. Both the bonding and antibonding orbitals are symmetrical about bond axis, *i.e.*, x-axis. These are, therefore , called σ-bonds. These two are denoted by σ $2p_z$ and σ* $2p_z$.

(iii) *Combination of* $2p_x$ *or* $2p_y$ *Orbitals :* If two $2p_x$ orbitals be considered, they can overlap side to side and not end to end as in the case of $2p_z$ orbitals.

The resultant molecular orbitals will therefore, he called p orbitals. The formation of bonding orbital, designated as π_x, 2p while the corresponding non-bonding orbital designated as σ* 2p.

Similarly, the formation of p, 2p and π_y^* 2p from the combination of two 2p, orbitals can be explained. The only difference is that the atomic and molecular orbitals lie now perpendicular to the plane of the paper.

Table 7 : Notations used to designate the various molecular orbitals are shown below

Molecular orbitals *Bonding*	*Antibonding*	*Atomic orbitals from which the molecular orbitals are formed*
σ 1*s*	σ*1s	1*s* orbitals of the two atoms
σ 2s	σ* 2s	2s orbitals of the two atoms
σ 2s	σ*2p	$2p_z$ orbitals of the two atoms
$\pi_x 2p$	$\sigma_x^* 2p$	2*px* orbitals of the two atoms
$\pi_y p$	$\sigma_y^* 2p$	2*py* orbitals of the two atoms

Differences between bonding and antibonding molecular orbitals: The differences between bonding and antibonding molecular orbitals may be summed up as follows :

Table 8

Bonding MO	*Antibonding MO*
1. A bonding MO gets formed by the addition-overlap of atomic orbitals.	An antibonding MO gets formed by the subtraction-overlap of atomic orbitals.
2. It is having lower energy than the AOs from which it is formed.	It is having higher energy than the AOs from which it is formed.
3. The electron charge density in between the nuclei is high and hence the repulsion between the nuclei is very low. This causes stabilisation of the bonding MO. In other words, the electrons in the bonding MO tend to favour stable bond formation.	The electron charge density in between the nuclei is very law and hence the repulsion between thenuclei is high. This causes de-stabilisation of the anti-bonding MO. In other words, the electrons in the antibonding MO tend to oppose bond formation.

Conditions for the formation of atomic orbitals : It is important to understand that any two atomic orbitals may not combine to form molecular orbitals may not combine to form molecular orbitals. In fact there are certain limitations to the combination of atomic orbitals which are summarised below :

(i) The energies of atomic orbitals must be similar in magnitude. According to this rule, a homonuclear ditomic molecule will not be formed if 1s-orbital of one atom overlaps with 2s-orbital of another atom. Similarly, a molecule will not be formed if 2s orbital of one atom overlaps with 2p-orbital of another atom because 1s, 2s and 2p-orbitals possess different energies. But in the case of formation of heteronuclear molecules, such combinations may be possible.

(ii) The atomic orbitals will combine only if they overlap to a considerable extent. The greater the overlap of atomic orbitals, the greater is the build-up of the charge between the nuclei. This rule is often referred to as the principle of maximum, overlap.

(iii) The atomic orbitals must have the same symmetry about the molecular axis. For example, p_x–orbital of one atom will not overlap with a s-orbital of another atom because the symmetry of s-orbital is not the same as that of p-orbital. In fact, the ++ overlap is cancelled by +– overlap. A p_z–orbital however, will combine with a s-orbital since both have the same symmetry.

A p_y–orbital will not combine with s-orbital as they possess different symmetry.

ELECTRONIC CONFIGURATIONS OF HOMONUCLEAR MOLECULES

With the help of the various principles discussed above, the electronic configurations f same of the common molecules may be written as below.

Hydrogen Molecule

When 1s orbitals from two hydrogen atoms overlap, σ 1s and σ* 1s molecular orbitals are formed. Since each hydrogen atom has only one electron, both electrons in the hydrogen molecule will be in the s 1s bonding orbitals, the antibonding orbital will remain vacant. The electronic configuration of the H_2 molecule is, therefore, $\sigma\ 1s^2$.

Number of covalent bonds in the molecule will be 1/2 (2 – 0) = 1. In other words, the two hydrogen atoms in the H2 molecule are bonded by a single covalent bond. In bonding molecular orbital σ 1s, there is an increase in electron charge density in the region between the two nuclei in hydrogen atoms. Thus, there is an accumulated negative charge between the two nuclei which holds these together at an equilibrium internuclear distance.

In antibonding molecular orbital, σ* 1s, there occurs a decrease in the charge density in the region between the two nuclei of hydrogen atoms. In the case, the negative charge is withdrawn from the region between the hydrogen nuclei and this increase the repulsive forces between the nuclei. In this region, the electrons repel each other. Thus, one can say that in antibonding molecular orbital, an electron spends half of its time on each of the individual hydrogen atom and consequently is never found in the region between the two hydrogen atoms. It means that the probability of finding the electron in the region between two hydrogen atoms in antibonding orbital is decreased.

In antibonding molecular orbital, there is a line between the two probability contours of hydrogen atoms. This is called a nodal plane, and at this nodal plane the electron charge density is zero.

Hydrogen Molecule Ion, H_2^+

This molecule has been detected spectroscopically when electric discharge is allowed to pass through hydrogen gas under reduced pressure. The molecule has only one electron and its electronic configuration is H_2^+ : σ $(1s)^1$ and *bond order* =1/2 (1 – 0) = 1/2. The molecular orbital energy level diagram for H_2^+ ion.

The positive value of bond order reveals that the bond does get formed and the molecule is stable. However, the bond in H_2^+ ion is weaker than that in H_2 molecule. The bond dissociation energy of H_2^+ ion (269 kJ mol–1) has been some what less than that of H_2 molecule (458 kJ $mole^{-1}$). The bond length in H_2^+ ion (1.04 Å) has been larger than that in H2 molecule (0.74 Å). These observations support the fact that bond in H_2^+ is weaker than the bond in H2 molecule.

Helium Molecule, He_2

Each helium atom is having two 1s–electrons and thus there are four electrons to get accommodated in He_2 molecule. As each molecular orbital, just like atomic orbital, can accommodate only two electrons of opposite spins, only one pair can be accommodated in the bonding s (1s) molecular orbital.

The other pair will have to be accommodated in the antibonding σ* (1s) MO.

The electronic configuration He_2 is, thus, He2 : σ $(1s)^2 \overset{*}{\sigma}$ (1s)

The zero bond order reveals that the molecule does not exist. Calculations have shown that *the antibonding effect of an antibonding MO is stronger than the bonding effect of a bonding MO. Consequently, there is no possibility whatsoever for the existence of* He_2 *molecule.* Thus, helium is monomatomic.

Lithium Molecule

Each lithium atom has an electronic configuration of $1s^2$ $2s^1$. We have a total six electrons from the two lithium atoms which are to be filled. Only the 2s electrons (valency electrons) are involved whereas the low lying $1s^2$ electrons do not take part in bonding. The molecular orbital energy diagram.

Its molecular orbital configuration is $[(\sigma\ 1s)^2\ (\sigma^*\ 1s)^2]\ [(\sigma\ 2s)^2]$.

Bond order = 1/2 [4 – 2] = 1 (one s bond). From the bond order, it is evident that Li_2 molecule should be formed. In fact, Li_2 species are present in sufficient concentrations at the boiling temperature of lithium.

Hence there is one Li – Li sigma bond. The bond dissociation energy of LiLi is appreciably low *i.e.*, 105kJ mol^{-1}. The bond length of the Li–Li is 2.67Å.

Comparison of Li–Li *and* H–H *bonds* :

(i) The sigma bond in Li–Li molecule is some what weaker than the sigma bond in H_2.

(ii) In lithium molecule, the outer electrons in $\sigma(2s)$ molecular orbital are shielded by the inner electrons. Hence the attraction between the nuclei and the electrons in $\sigma(2s)$ molecular orbital is less.

Beryllium Molecule, Be_2

Each of the beryllium atom possesses electronic configuration $1s^2$ $2s^2$. So there is a total of eight electrons to be filled in four molecular orbitals. The configuration is $[\sigma\ 1s)^2\ (\sigma^*\ 1s)^2]\ [(\sigma\ 2s)^2]$.

The bond order is 1/2 (4 – 4) = 0.

This indicates that molecular beryllium will not exist. This is in accordance with the experimental results.

Nitrogen Molecule

Electronic configuration of the nitrogen atom is 1s2 2s2 $2p_x^1$ $2p_y^1$. When the two atomic orbitals from two nitrogen atoms combine, there will be 4 electrons in the N_2 molecule and these will be distributed themselves in the various molecular orbitals as follows.

$$[(\sigma\ 1s)^2\ (\sigma^*\ 1s)]\ [(\sigma\ 2s)^2\ (\sigma^*\ 2s)^2\ (\sigma\ 2p)^2\ (\pi_x\ 2p)^2\ (\sigma_y\ 2p)^2].$$

The three antibonding orbitals biz. $\pi_x^* 2p$, $\pi_y^* 2p$ and $\sigma^* 2p$ remain vacant, σ 1s and σ* 1s orbitals do not take part in bond formation and are known as non-bonding orbitals.

$(\sigma\ 2s)^2$, $(\sigma\ 2p)^2$, $(\pi_x\ 2p)^2$ and $(py\ 2p)^2$ from bonding orbitals while (s* 2s)2 forms antibonding orbitals. Therefore, number of bonding electron is 8 while the number of antibonding electrons is 2 only. Thus, the number of covalent bonds is 1/2 (8 – 2) = 3. This means that two nitrogen atoms in the N_2 molecule are bonded by a triple covalent bond. The formation of the nitrogen molecule.

In the formation of N_2 molecule only one antibonding orbital is involved while the remaining four are all bonding orbitals. Hence the nitrogen molecule is highly stable. This is confirmed by its high bond dissociation energy viz., 945.6 kJ mol^{-1} and small bond length equal to 1.10Å. Also N_2 molecule is diamagnetic because there are no unpaired electrons in the molecule.

Oxygen Molecule

The atomic number of oxygen atoms is eight. The electronic configuration of the oxygen atoms is $1s^2\ 2s^2\ 2p_x^2\ p_y^2\ 2p_z^2$. When the two oxygen atoms combine, there will be 16 electrons in the O_2 molecule and these will be distributed in the various molecular orbitals as follows :

$[(\sigma\ 1s)^2\ (\sigma^*\ 1s)^2]\ [(\sigma\ 2s)^2\ (\sigma^*\ 2s)^2\ [\sigma\ 2p)^2\ (\pi_x\ 2p)^2\ (\sigma_y\ 2p)^2\ (\pi_x^*\ 2p)^1$ $(\pi_y^*\ 2p)^1$. $[(\sigma\ 1s)^2$ and $(\sigma^*\ 1s)^2]$ do not take part in the chemical bonding because orbitals containing valency electrons take part only. $(\sigma\ 2s)^2$ $(\sigma\ 2p)^2\ (\pi_x\ 2\ p)^2\ (p_y\ 2p)^2$ are bonding orbitals and, thus, they contribute eight electrons. $(\sigma^*\ 2s)^2$, $(\pi_x^*\ 2p)^1$ and $(\pi_y^*\ 2p)^1$ are antibonding orbitals and they contribute 2 + 1 + 1, *i.e.*, 4 electrons.

Thus, the number of covalent bonds will be 1/2 (8 – 4) = 2, *i.e.*, the two oxygen atoms in O_2 molecule are bonded by a double bond. The oxygen molecule is represented. Due to the presence of four electrons in anitbonding MOs oxygen molecule is less stable than nitrogen molecule. Its bond dissociation energy is much less *i.e.*, 494.6 kJ mol^{-1}, and bond length is larger *i.e.*, 1.21Å.

O_2^- **–ion**

This type of ion, also called the superoxide ion is present in potassium superoxide, KO_2. This ion has seventeen electrons. The structure of this ion is related to that of the oxygen (O_2) molecule. The ion possesses one extra electron which occupies either $\pi_x^* 2p$ or $\pi_y^* 2p$. Thus,

$[(\sigma\ 1s)^2 (\sigma^*\ 1s)^2]\ [(\sigma\ 2s)^2 (\sigma^*\ 2s)^2 (\sigma\ 2s)^2 (\pi_x\ 2p)^2 (\pi_y\ 2p)^2 ((\pi_x^*\ 2p)^2 (\pi_y^*\ 2p)^1\ ]$.

The $(\sigma\ 1s)^2$ and $(\sigma^*\ 1s)^2$ and do not take part in the chemical bonding. The $(\sigma\ 2s)^2$ $(\sigma\ 2p)^2$, $(\pi\ 2p)^2$ and $(\pi_y\ 2p)^2$ are bonding orbitals and they contribute 8 electrons. The $(\sigma^*\ 2s)$, $(\pi_x^*\ 2p)^2$ and $(\pi_y^*\ 2p)^1$ are antibonding orbitals and they contribute 2 + 2 + 1 or 5 electrons. Thus, the number of covalent bonds will be 1/2 (8 – 5) or 1.5 *i.e.*, the bond order is 1.5.

O_2^{2-} **Ion**

This type of ion, also called the peroxide ion, is present in sodium peroxide, NaO_2. This ion contains 18 electrons which are arranged in the following way :

$[(\sigma\ 1s)^2 (\sigma^*\ 1s)^2]\ [(\sigma\ 2s)^2 (\sigma\ 2p)^2 (\pi_x\ 2p)^2 (\pi_y\ 2p)^2 (\pi_x^*\ 2p)^2\ (\pi_y^*\ 2p)^2\ ]$

The bonding and antibonding p orbitals completely cancel each other, leaving one sigma bond from the filled σ 2p orbital and a bond order of one.

Fluorine Molecule

The atomic number of fluorine is nine. It has electronic configuration $1s^2\ 2s^2\ 2p^5$. When fluorine molecule is formed, there will be 18 electrons in the F_2 molecule and these will be distributed in various molecular orbitals as :

$[(\sigma\ 1s)^2 (\sigma^*\ 1s)^2]\ [(\sigma\ 2s)^2 (\sigma^*\ 2s)^2 (\sigma\ 2p)^2 (\pi_x\ 2p)^2 (\pi_y\ 2p)^2 (\pi_x^*\ 2p)^2 (\pi_y^*\ 2p)^2\ ]$ $(\sigma\ 1s)^2$ and $(\sigma^*\ 1s)^2$ do not take part in bonding because only electrons in the valency shell take part. $(\sigma\ 2s)^2\ (\sigma\ 2p)^2\ (\pi_x\ 2p)^2\ (\pi_y\ 2p)$ are bonding orbitals and contribute 8 electrons. $(\pi_x^*\ 2p)^2, (\pi_y^*\ 2p)^2$ and $(\sigma^*\ 2s)^2$ form antibonding orbitals and they contribute, therefore, 6 electrons. Thus, the number of covalent bons is 1/2 (8 – 6) = 1. Thus, the bond between two fluorine atoms is a single bond.

As all the electrons in bonding as well as in anitbonding orbitals are completely paired, it means that fluorine molecule is diamagnetic. Similarly structures of other halogens (Cl, Br and I) can be explained.

Neon Molecule

Its structure is similar to helium. In the formation of Ne_2, if possible sixteen outer electrons have to be accommodated in molecular orbitals. The electronic configuration of hypothetical Na_2 molecule is

$$[(\sigma\ 1s)^2\ (\sigma^*\ 1s)^2]\ [(\sigma\ 2s)^2\ (\sigma^*\ 2s)^2\ (\sigma\ 2p)^2$$
$$(\pi_x\ 2p)^2\ (\pi_y\ 2p)^2 (\pi_x^*\ 2p)^2\ \ (\pi_y^*\ 2p)^2\ ]\ (\sigma^*\ 2p)$$

Bond order in this case is (10 – 10) = 0. As antibonding MOs slightly dominate over the bonding MOs, N_2 molecule would be incapable of existence.

Bond Strength : From the above examples, it is evident that bond strength decreases with the increase in the number of antibonding orbitals. As nitrogen molecule does not involve antibonding orbitals (only valence orbitals are to be considered) whereas oxygen molecule does, it means that nitrogen is more stable than oxygen molecule. Similarly, the stability of other molecules can be compared on the basis of number of antibonding orbitals used in their formation.

Energy Level Diagram for Molecular Orbitals : It is known that Is atomic orbitals of two atoms from two MOs designated as s (1s) and $\overset{*}{\sigma}$ and a (1s). The 2*s* and 2*p* atomic orbitals (eight AOs on two atoms) result in the formation of the following eight MOs :

Bonding MOs : s (2s) s $(2p_z)$ $\pi\ (2p_x)$ $\pi\ (2p_y)$

Antibonding MOs : $\overset{*}{\sigma}\ (2s)$ $\overset{*}{\sigma}\ (2p_z)$ $\overset{*}{\pi}\ (2p_x)$ $\overset{*}{\pi}\ (2p_y)$

Note : *The d-orbitals are not included in the present discussion since the purpose is to consider the formation of simple diatomic molecules only.

The energy levels of different molecular orbitals. In this diagram, the molecular orbitals have been shown in the centre and the atomic orbitals which combine have shown on the two extremes.

The atomic orbitals of same energy have been shown at the same level while the bonding MOs are shown lower and the antibonding MOs have been shown higher than the corresponding atomic orbitals. This indicates that while the bonding MOs are more stable, the antibonding MOs are less stable than the AOs from which they are formed. It is evident from the figure that *the extent of stabilization for a bonding MO is always equal to the extent of destabilisation for the corresponding antibonding MO.*

However, from spectroscopic studies it has been found that the above sequence of energy levels of MOs is not correct for all molecules. It has been observed that in case of homonuclear diatomic molecules of second row elements of the periodic table upto N_2 (viz. Li_2 Be_2, B_2, C_2, N_2), s ($2p_z$) *MO has been found to be higher in energy than the p (2px) and p* ($2p_y$) *MOs.* The reversal in the sequence of energy levels of MOs is ascribed to the positively of *mixing of 2s and $2p_z$ atomic orbitals,* as discussed below.

MIXING OF ORBITALS

It has been known that s (2s) and $\overset{*}{\sigma}$(2s) MOs get formed by the combination of 2s atomic orbitals of two atoms while σ ($2p_z$) and $\overset{*}{\sigma}$ ($2p_z$) MOs get formed by the combination of $2p_z$ atomic orbitals of two atoms. However, if the energy differences between 2s and 2p atomic orbitals are small, there is a possibility of the *mixing of these orbitals* (cf. hybridisation of atomic orbitals) because of which the σ (2s) and $\overset{*}{\sigma}$ (2s) MOs do not retain pure s character and, similarly, the σ ($2p_z$) and $\overset{*}{\sigma}$ ($2p_z$) MOs do not retain pure p character.

On the other hand, *all the four orbitals acquire s mixed sp character.* Because of this s-p mixing, the energies of all the four orbitals change in such a way that the MOs s (2s) and $\overset{*}{\sigma}$ (2s) become more stable and thus get lowered in energy whereas the MOs σ($2p_z$) and $\overset{*}{\sigma}$ become less stable and are thus raised in energy.

As π (p) orbitals are not involved in mixing, the energies of p ($2p_x$) and p ($2p_y$) and p ($2p_y$) MOs remain unchanged.

The s (2pz) *MO is raised to such an extent that it becomes higher in energy than p (2px) and p (2py) MOs.* The new sequence of energy of MOs, therefore, becomes as follows :

$$\sigma(1s), \overset{*}{\sigma}(1s), \sigma(2s), \overset{*}{\sigma}(2s), \pi(2p_x) = \pi(2p_y), \sigma(2p_z), \overset{*}{\pi}(2p_y), \overset{*}{\sigma}(2p_z)$$

—— Energy increase ⟶

The energy levels of different molecular orbitals obtained because of mixing of 2s and $2p_z$ atomic orbitals.

Now, a question arises : How is it that the mixing of 2s and 2p atomic orbitals take place only in case of second row elements upto nitrogen only and not in case of oxygen and fluorine. The answer lies in the differences in energies of 2s and 2p atomic orbitals of atoms of second row elements, as depicted in the following Table 9.

Table 9 : Energy differences of 2s and $2p_z$ AOs of Atoms of Second Row Elements

Atoms	*Li*	*Be*	*B*	*C*	*N*	*O*	*F*
Difference in energies of 2s and $2p_z$ AOs (kJ $mole^{-1}$)	178	262	449	510	570	1430	1970

It is evident from the above table that while the differences in energies of 2s and $2p_z$ orbitals for atoms up to nitrogen are small, these differences becomes significantly large for oxygen and fluorine. Consequently, while the mixing of 2s and $2p_z$ AOs becomes readily possible in case of elements upto nitrogen, such a mixing in case of oxygen and fluorine has been practically ruled out.

It can be easily visualised that die to large energy differences, mixing of 1s and 2s atomic orbitals will also not take place. It is to be noted that out of the three 2p orbitals, only the $2p_z$ AO can combine with 2s atomic orbital due to symmetry considerations.

Filling up of Molecular Orbitals

The principles involved in filling up of electrons in molecular orbitals may sum up as below.

(i) The molecular orbitals with the lowest energy is filled first. Thus, s (1s) is filled first whereas s* (2p) is filled in the last. The order of filling up of molecular orbitals is as follows :

$$\sigma\ (1s) < \sigma^*\ (1s) < \sigma\ (2s)\ \sigma^*\ (2s) < \sigma\ (2p)$$

$$< \begin{pmatrix} \pi_x(2p) \\ \pi_y(2p) \end{pmatrix} < \begin{pmatrix} \pi_x^*(2p) \\ \pi_y^*(2p) \end{pmatrix} < \sigma^*\ (2p)$$

(ii) The maximum number of electrons in a molecular orbitals cannot exceed two,

(iii) The molecular orbitals are first singly filled up and paring starts only when more electrons are to be accommodated (*Hund's Rule*). Non-bonding orbitals are filled up just like bonding orbitals.

(iv) If there are two molecular orbitals at the same energy level, electrons will occupy separate orbitals in order to avoid each other, thus, leading to unpaired electron spins.

(v) *Bond order :* As an electron in an antibonding molecular orbitals cancels out the stability introduced by the electron in a bonding molecular orbital, it means that in order for bonding of atoms to occur there should be an excess of bonding electrons over antibonding electrons. In case where *the number of bonding and antibonding electrons are equal, no bond will be formed between the atoms.* However, the number of covalent bonds are related to bonding and antibonding electrons by a term called bond order which is defined as :

$$\text{Bond order} = \frac{\text{No. of electrons in bonding orbitals} - \text{No. of electrons in antibonding orbitals}}{2}$$

ELECTRONIC CONFIGURATIONS OF HETERONUCLEAR DIATOMIC MOLECULES

Let us now consider the cases of heteronuclear diatomic molecules, in which the two atoms constituting a molecule are different. The principles involved in the distribution of electrons remain the same as discussed before.

However the molecular orbital diagram will not be symmetrical as in case of homonuclear diatomic molecules. Because of the difference in the electronegativites of the atoms, the electrons in bonding molecular orbital tend to spend more time near the more electronegative atom. On the other hand, the electrons in antibonding molecular orbital remain closer to the less electronegative atom.

Let us now discuss the electronic configurations and molecular orbital diagrams for some common hetero–nuclear diatomic molecules.

Carbon Monoxide Molecule, CO

The electronic configurations of C and O atoms are $1s^2\ 2s^2p^2$ and $1s^2\ 2s^2p^4$ respectively. Thus, there are 4 electrons in the outer shell of carbon and 6 electrons in the outer shell of oxygen. Hence, a total of 10 outer electrons have to be accommodated in the molecular orbitals of CO molecule. Because of higher electronegativity of oxygen, its atomic orbitals would be of lower energy than the corresponding atomic orbitals of carbon. Because of this energy difference, the bonding and antibonding molecular orbitals will get more contribution from atomic orbitals of lower energy, *i.e.*, the atomic orbitals of oxygen and would be closer to it in energy than the antibonding molecular orbitals which would be closer to carbon in energy. The bonding molecular orbitals would be having more characteristics of atomic orbitals of oxygen and antibonding molecular orbitals would have more characteristics of carbon.

The electronic configuration of CO molecule is.

$$[(\sigma\ 1s)^2\ (\sigma^*\ 1s)^2]\ \sigma\ (2s)^2\ \sigma^*\ (2\ s)^2\ \pi\ (2p_x)^2\ \pi\ (p_y)^2\ \sigma\ (2p_z)^2.$$

Bond order in this case is 1/2 (8 – 2) = 3. Thus, CO contains a triple bond *i.e.*, one sigma and two p bonds. Its bond dissociation energy has been found to be 1067 kJ mol^{-1} and bond length equal to 1.14Å..

From the molecular orbital energy level diagram, the following facts may be noted :

(i) We know that the electronegativity of oxygen is greater than that of carbon. Therefore, the 2s and 2p orbitals have been placed lower than the 2s and 2p orbitals of carbon.

(iii) We also know that CO is isoelectronic with N_2. Therefore, it has the same molecular orbital configuration as N_2 molecule has.

Hydrogen Fluoride Molecule, HF

The electronic configuration of hydrogen and fluorine have been $1s^1$ and $1s^2\ 2s^2\ p^5$, respectively. In the formation of HF molecule, only the

2p–electrons of fluorine atom would combine effectively with the solitary electron of hydrogen atom. It is known that only a p_z orbital is able to combine with an s-orbital. The combination of a p_x *or* p_y orbital with an s-orbital is ruled out on symmetry considerations. The $2p_z$ orbital of fluorine yields a highly effective overlap with the s-orbital of hydrogen to form the σ (sp) molecular orbital.

It is known that the σ *(sp) is the only MO formed in HF molecule.* This is, evidently, a bonding orbital which is fully filled with 2 electrons. The rest of the electrons remain in their atomic orbitals. The MO formed in the case of HF molecule will not be symmetrical. The asymmetry occurs because the energies of H (1s) and F ($2p_z$) atomic orbitals are not the same.

It can be seen that the bonding MO has an energy which lies closer to that of $2p_z$ orbitals of fluorine than to that of higher energy 1s–orbital of hydrogen.

Hydrogen Chloride Molecule HCl

The electronic configurations of H and Cl atoms have been

$1s^1$ and $1s^2\ 2s^2p^6\ 3s^2p^5$, respectively. In this case also, there is only one bonding MO and one anti-bonding MO formed by the combination of 1s–orbital of hydrogen with $3p_z$ orbital of chlorine. The rest of the electrons of the chlorine atom remain in their respective AOs.

The electronic configuration of *HCl* is thus represented as follows :

$$\text{HCl : KL } (3s)^2\ \sigma(sp)^2\ (3p_x)^2\ (3p_y)^2$$

In this case also, the MO formed would not be symmetrical. The MO diagram of HCl molecule will be exactly similar. 5.64 for HF molecule.

Similarities Between Valence Bond and Molecular Orbital Methods

(i) When a bond is formed, both methods envisage an increase in charge density between the nuclei.

(ii) In both methods it is assumed that bond is only formed if orbitals of bond atoms overlap and have appropriate symmetry about the molecular axis.

(iii) In both methods, Aufbau's principle (Fault's principle and Hand's rule) is followed in filling atomic as well as molecular orbitals.

(iv) Both theories account for directional nature of covalent bonds.

Table 10 : Differences between molecular orbital and valence bond methods

Molecular Orbital Method	*Valence Bond Method*
1. In molecular orbital theory a molecule is composed of such atoms which lose their individual character.	1. In valence bond method, a molecule is composed of such atoms which do not lose their individual character when linked to other atoms.
2. An electron moves in the field of more than one nucleus, *i.e.*, the molecular orbitals are polycentric.	2. An electron moves in the field of one nucleus only, *i.e.*, atomic orbitals are monocentric.
3. When dealing with energy levels of electrons in molecules, electrons can be dealt with individually.	3. When dealing with energy levels of electrons in molecules in the valence bond method, electrons are dealt in pairs.
4. Molecular orbital theory accepts the ionic structures like $A^+ B^-$ and $A^- B^+$ in a molecule like AB.	4. Valence bond method rejects the ionic structures like $A^+ B^-$ and $A^- B^+$ in a molecule like AB.
5. Molecular orbital theory is more convenient for description of exited state in molecules.	5. It does not offer simple methods.
6. This theory readily explains the paramagnetism of the oxygen and sulphur molecules.	6. It does not readily explain.

From the above examples, it follows that there occurs a change in bond energy and bond length in the diatomic molecules. The data are listed in Table 11.

Table 11 : Characteristics of diatomic molecules.

Molecule	Li_2	B_2	C_2	N_2	O_2	F_2
Number of bonding electron	2	2	4	6	4	2
Bonding energy (kJ/mole)	105	290	630	945	496	151
Bonding length, Å	2.67	1.59	1.31	1.10	1.21	1.42

United Atom Method

In the LCAO method, atoms are brought from infinite distance gradually near one another till they take up a position of equilibrium in one molecule. Thus, the individual atomic orbitals yield molecular orbitals which may be bonding if the spins are paired or antibonding if the spins are unpaired.

In the united atom method, we start with an united atom in which the nuclei are superimposed over one another. Then, the individual nuclei are gradually separated till they acquire a position of equilibrium. It is important to remember that the united atom is purely hypothetical and has no actual existence.

Let us illustrate the formation of hydrogen molecule by the united atom method. In this treatment, we consider the structure of H_2 where the hydrogen nuclei are superimposed over each.

In this structure there are no inter–atomic forces of attraction and repulsion. But there are naturally a great deal of repulsion between the two nuclei and electrons. Thus, the system will have higher energy and it will not be stable. If we now separate the nuclei very slowly, the repulsive forces will decrease and the energy of the system will be lowered. At equilibrium position, the two nuclei will be at a distance at which the forces of repulsion are minium. This is also the position at which the energy of the system is lowest. This is the position at which the molecular orbital is formed. If we remove the atoms to infinity, the energy of individual atomic orbitals increase. Thus, we can represent the relationship between the orbitals in the united atom, molecular orbitals when the atoms are at equilibrium distance, and the individual atomic orbitals when atoms are at infinity as follows :

Oribital of united atom ⟶ Molecular orbitals formed when atoms are at equilibrium position

⟶ Atomic orbitals of individual atoms at inf inity

From the above, it is evident that molecular orbitals are intermediate in position between orbitals of united atom and orbitals of the separated atom.

Mulliken correlation diagram for atoms which from a diatomic molecule.

Coulson's Mot Treatment of Co Molecule

According to coulson, the molecular diagram of CO given above is not correct and should involve the concept of hybridisation.

The electronic configuration of carbon a toms is $2s^2\,2p_x^0\ 2p_y^1\ 2p_z^1$. The 2s and 2px orbitals undergo hybridisation to form two sp–hybrid orbitals which are represented as (sp). Thus, after hybridisation, the electronic configuration of carbon becomes as follows :

$$(sp)_c^2\ (sp)_c^1\ 2p_x^0\ \ 2p_y^0$$

Also 2s and $2p_z$ atomic orbitals on oxygen atom ($2s^2, 2p_x^2\ 2y_y^1\ \ 2p_z^1$) undergo hybridisation to form two sp hybrid orbitals and the electronic configuration of oxygen atom then becomes as follows :

$$(sp)_0^2\ (sp)_0^1\ 2p_x^2\ \ 2p_y^2$$

As electron pairs $(sp)_0^1$ and $(sp)_0$ hybrid orbitals do not participate in bonding, these are known as σ–non bonding electron pairs (σ_{ab}– pairs). Thus, the CO molecule is having two lone pairs. one pair on $(sp)_c$ and $(sp)_0$. As $(sp)_c$ hybrid orbital having a lone pair has higher energy than that having a single electron, it means that it has a higher degree of p–character and is very reactive. This explains the properties of CO. Also $(sp)_c$ points away from O–atom.

As $(sp)_0$ hybrid orbital having lone pair has lower having lone pair has lower energy than that having a single electron, it has a high degree of s–orbital character and hence is very stable *i.e.* unreactive, hybrid orbital.

Now there remain two singly filed $(sp)_c$ and $(sp)_0$ hybrid orbitals and two 2p–orbitals ($2p_x$ and $2p_y$) on both the atoms which overlap to form the following orbitals.

(i) n $(s\ p)_c$ and $(sp)_0$ overlap to form $\sigma_{sp}{}^+$ and $\pi_{sp}{}^*$ orbitals.

(ii) $2p_x$ atomic orbital on carbon atom and $2p_x$ atomic orbital on oxygen atom overlap to form $p_x{}^h$ and $p_x{}^*$ molecular orbitals.

(iii) $2p_y$ atomic orbital on carbon atom and $2p_y$ atomic orbital on oxygen atom overlap to form $\pi_y{}^b$ and $\pi_y{}^*$ molecular orbitals.

The relative order of energy of these molecular orbitals and sp hybrid orbitals having lone pairs on each atom is as follows.

$$(sp)_0 < \sigma_{sp}{}^b < \pi_x{}^b = \pi_y{}^b < (sp)_c < \pi_x{}^*$$
$$= \pi_y{}^* < \sigma_{sp}{}^*$$

Therefore, the molecular orbital configuration of CO is as follows :

$$(sp)0^2, (s_{sp}{}^b)^2, (p_x{}^b)^2 = p_y{}^b)^2, (sp)_c{}^2, (p_x{}^*)^0,$$
$$= (\pi_y{}^*)^0 (\sigma_{sp}{}^*)^0$$

The bond order of CO is 1/2 (6 – 0) or 3. This shows that carbon and oxygen in CO molecule are linked together by a triple bond (1s + 2p bonds).

Nitrosyl ion (N O) : The NO^+ ion has ten electrons in its valence shell. Therefore, it is isoelectronic with N_2 molecule which has also ten electrons. Thus, the NO^+ ion has the same molecular orbital configuration as N_2 molecule, *i.e.*,

$$(\sigma_{2s}{}^b) (\sigma_{2s}{}^*) (\pi_x{}^b)^2 = (\pi_y{}^b)^2, (\sigma_z{}^b)^2.$$

Nitric oxide molecule, NO : The electronic configuration of N and O atoms have been $1s^2\ 2s^2\ p^3$ and $1s^2\ 2s^2\ p^4$, respectively. There are 5 electrons in the outer shell of nitrogen and 6 electrons in the outer shell of oxygen. Thus, a total of 11 outer electrons have to be accommodated in the molecular orbitals of NO molecule.

The electronic configuration of NO molecule is, thus as follows :

$$(\sigma\ ls)^2 (\sigma^*\ 1s)^2\ \sigma\ (2\ s)^2\ \sigma^*\ (2\ \sigma)^2\ \sigma\ (2p_z)^2\ \pi\ (2p_x)^2\ \pi\ (2p_y)^2\ \pi^*(2p_x)^1.$$

The bond order in this case is $\frac{1}{2}$ (8 – 3) = $2\frac{1}{2}$. This represents one σ–bond and two π–bonds less the antibonding effect of one unpaired electron in the $\pi^*\ (2p_x)$ molecular orbital, the presence of this antibonding electron makes the NO molecule less stable than the N_2 molecule. This has been confirmed by the fact that the bond dissociation energy of nitric oxide, viz. .667.8 kJ mol^{-1} has been less than that of nitrogen which is 945.6 kJ mol^{-1}.

As NO contains an unpaired electron, it is *paramagnetic.*

As in the case of CO molecule, the bonding molecular orbital would be having greater characteristics of atomic orbitals of oxygen and the antibonding molecular orbitals would have greater characteristics of atomic orbitals of nitrogen. The bond order 2½ of nitric oxide indicates that this molecule is paramagnetic.

4

Linnett Theory and Resonance

INTRODUCTION

The Linnett theory is applied to many molecules, it is found that it becomes unnecessary to take into account the concept of resonance. We will now discuss some examples which will illustrate this fact.

Sulphur dioxide : The resonance structures of SO_2 are as follows:

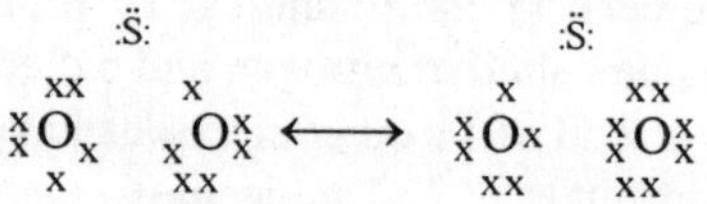

I II

The above two structures differ only in the position of the double bond. Two S–O bonds are equivalent in the resonance hybrid of the structures (I) and (II). According to Linnett, sulphur dioxide is represented by a single structures. In the above structure, the resonance is also shown. Thus, if the Linnett theory is followed, it is possible to show resonance with a single structure. It means that the Linnett theory does not completely eliminate resonance.

ADVANTAGES OF LINNETT DOUBLE QUARTET THEORY

(i) This theory is also known as non-pairing spatial orbital method. This theory is intermediate between the valence bond theory and the molecular orbital theory.

(ii) This theory does not lay a great emphasis that electrons should be paired if they are to form bonds as in the molecular orbital theory. In this respect it is superior to the molecular orbital theory because of the allowance it makes for charge repulsion effects, its ability to predict the shapes of the molecules and its expansibility to complex system.

(iii) Linnett theory is superior to valence bond theory because it explains bonding in many paramagnetic molecules and because of its ability to represent a molecule by a single acceptable structure rather than several contributing structures. However, Linnett's theory shows resonance with a single structure in some compounds. Therefore, it does not discard resonance altogether.

(iv) It also explains why NO does not exist as $(NO)_2$.

(v) When quantitative calculations are done for the Linnett's theory, they are found to be superior over the simple valence bond theory and molecular orbital theory.

Bent Rule

This rule may be stated as follows :

More electronegative substituents "prefer" hybrid orbitals having less s character and more electropositive substituents "prefer" hybrid orbitals having more s character. Although proposed as an empirical rule, in the chlorofluorides of phosphorus, it is substantiated by molecular orbital calculations.

An example of Bent's rule is provided by the fluoromethanes. In CH_2F_2 the F–C–F bond angle is less than 109½°, indicating less than 25% s character, but the H–C–H bond angle is larger and the C–H bond has more s character. The bond angles in the other fluoromethanes yield similar results.

The tendency of more elctronegative substituents to seek out the low electronegativity $p_z d_z^2$ apical orbital in TBP structures is often termed "apicophilicity." It is well illustrated in a series of oxysulfuranes of the type prepared by *Martin* and *Coworkers.* These, as well as related phosphoranes, provide interesting insight into certain molecular rearrangements. Bent's rule is also consistent with and may provide alternative rationalization for Gillespie's VSEPR model. Thus, the Bent's rule prediction that highly electronegative substituents will "attract" *p* character and reduce bond angles is compatible with the reduction in angular volume of the bonding pair when held tightly by an electronegative substituent. Strong, s-rich covalent bonds require a larger volume in which to bond.

Thus, double bonded oxygen, despite the high electronegativity of oxygen, seeks *s*- rich orbitals because of the shortness and better overlap of the double bond. Again, the explanation, whether in purely s-character terms (Bent's rule) or in larger angular volume for a double bond (VSEPR), predicts the correct structure. It is sometimes philosophically unsettling to have multiple "explanations" when we are trying to understand what makes molecules behave as they do; it is only human "to want to know for sure" why things happen. On the other hand, alternative nonconflicting hypotheses give us additional ways of remembering and predicting facts, and if we seem to find them in conflict, we have made either. (1) a mistake (which is good to catch) or (2) a discovery.

The mechanism operating behind Bent's rule is not completely clear. One factor favouring increased *p* character in electronegative substituents is the decreased bond angles of *p* orbitals and the decreased steric requirements of electronegative substituents. There may also be an optimum "strategy" of bonding for a molecule in which the character (and hence improved overlap) is concentrated in those bonds in which the electronegativity difference is small and covalent bonding is important. The *p* character, if any, is then directed towards bonds to electronegative groups. The latter will result in greater ionic bonding in a situation in which covalent bonding would be low anyway (because of electronegativity differences).

Some light may be shed on the workings of Bent's rule by observations of apparent exceptions to it. The rare exceptions to broadly useful rules are unfortunate with respect to the universal application of those rules. They also have the annoying tendency to be confusing to someone who is encountering the rule for the first time. On the other hand, any such exception or apparent exception is a boon to the research scientist since it almost always provides insight into the mechanism operating behind the rule. The phosphorus atom is in an approximately tetrahedral environment using four σ bonds of approximately sp^3 character. We should expect the more electronegative oxygen atoms to bond to s-poor orbitals on the phosphorus and the two oxygen atoms in the ring do attract hybridization of about 20% s.

The most electropositive substituent on the phosphorous is the bromine atom and Bent's rule would predict an s-rich orbital, but instead it draws another s-poor orbital, slightly less than 20% s. The only s-rich orbital on the phosphorus atom is that involved in the σ bond to the exocyclic oxygen. This orbital has nearly 40% s character. This oxygen ought to be about as electronegative as the other two, so why the difference? The answer probably lies in the overlap aspect :

(1) The large bromine atom has diffuse orbitals that overlap poorly with the relatively small phosphorus atom ; thus, even though the bromine is less electronegative than the oxygen, it probably does not form as strong a covalent bond.

(2) The presence of a π bond shortens the exocyclic double bond and increases the overlap of the a orbitals. If molecules respond to increase in overlap by rehybridization in order to profit from it, the increased s character then becomes reasonable. From this point of view, Bent's rule might be rewarded : *The p character tends to concentrate in orbitals with weak covalency (from either electronegativity or overlap considerations), and s character tends to concentrate in orbitals with strong covalency (matched electronegativities and good overlap).*

Some quantitative support for the above qualitative arguments comes from average bond energies of phosphorus, bromine, and oxygen.

$$\text{P—Br } 264 \text{ kJ mol}^{-1}$$

$$\text{P—O } 335 \text{ kJ mol}^{-1}$$

$$\text{P} = \text{O } 544 \text{ kJ mol}^{-1}$$

Bent's rule is a useful tool in inorganic and organic chemistry. For example, it has been used to supplement the VSEPR interpretation of the structures of various nonmetal fluorides, and should be applicable to a wide range of questions on molecular structure.

BERYLLIUM FLUORIDE

The atomic number of beryllium is 4. Its electronic configuration is $1s^2\ 2s^2$. As no unpaired electrons are available, the formation of BeF_2 is not expected to be formed. But BeF_2 is known. In order to explain its formation, it is assumed that one of the 2s electrons becomes unpaired and gets excited to 2p-orbital. Now two unpaired electrons are available in the excited state of beryllium and, therefore, it can form BeF_2 by sharing its two, electrons with two electrons from two fluorine atoms. But formation of BeF_2 involves one s and one p-orbital. Therefore, the two bonds should be of different strength. But actually these are of equal strength and the molecule BeF_2 is linear with the bond angle F–Be–F equal to 180°. However, it is explained by hybridisation.

$_4$Be (ground state) = $1s^2\ 2s^2$ (Fig. 19 A)

$_4$Be (Excited state) = $1s^2\ 2s^1\ 2p^1$ (Fig. 19 B)

$_9F = 1s^2\ 2s^2\ 2p_x^{\ 2}\ 2p_y^{\ 2}\ 2p_z^{\ 1}$

The energy of one *s*- and one p-orbital is combined and two *sp-hybrid* orbitals of the same shape and energy are obtained. These two sp-hybrid orbitals overlap with p-orbitals of two fluorine atoms to form BeF_2.

In BeF_2, the two hybrid orbitals must lie as far apart from each other as possible in order to minimise the force of repulsion and thus to have a stable structure. Hence, the two hybrid orbitals point in opposite directions. The molecule will therefore be linear and the bond angle F–Be–F will be equal to 180°.

Thus, hybridisation in BeF_2 explains that :

(i) Two bonds are of equal strength, and

(ii) The shape of the molecule is linear.

Hybridisation of sp^2 type : *The combination of one s and two p-orbitals to form three orbitals of equal energy is known as sp^2 hybridisation.* These three bonds are co-planar and directed at angle of 120° to each other.

Properties of sp^2 Hybrid Orbitals

(i) All the three *sp^2* hybrid orbitals are completely equivalent in energy and shape.

(ii) The relative overlapping power of each hybrid orbital is 1.99.

(iii) All the three *sp^2* hybrids lie in the plane of p-orbital and are oriented at 120° with respect to each other.

(iv) Theoretically, it has been deduced that the ratio of p-character to s-character is 2 : 1 in each of sp^2 hybrids.

(v) The three *s^2* hybrids have their maximum overlapping along one of the three axes. The three hybrids are denoted by $sp_a^{\ 2}$, $sp_b^{\ 2}$ and $sp_c^{\ 2}$. The normalised wave function for these hybrids are as follows :

$$\psi = (sp_a^{\,2}) = \frac{1}{\sqrt{(3)}}\ \psi\,(s) + \sqrt{\left(\frac{2}{3}\right)}\ \psi\,(p)$$

$$\psi = (sp_b^{\,2}) = \frac{1}{\sqrt{(3)}}\ \psi\,(s) - \frac{1}{\sqrt{(6)}}\ \psi\,(p) + \frac{1}{\sqrt{(2)}}\ \psi\,(p)$$

$$\psi\,(sp_c^{\,2}) = \frac{1}{\sqrt{(3)}}\ \psi\,(s) - \frac{1}{\sqrt{(6)}}\ \psi\,(p) - \frac{1}{\sqrt{(2)}}\ \psi\,(p)$$

From the above equations it follows that the ration of p-character to s-character is 2 : 1 in each of the hybrid wave functions.

Example : An interesting example to illustrate the sp^2 hybridisation is the formation of BF_3.

The boron atom is the central atom in BF_3 and its electronic configuration is $1s^2\, 2s^2\, 2p^1$. The atom in its ground state has one unpaired electron so that it can form one covalent bond but in the excited state, there are three unpaired electrons and hence three bonds of unequal strength can be formed with three fluorine atoms.

But all bonds are of equal strength. In order to explain this, it is assumed that one s and two p-orbitals of excited boron atom undergo hybridisation to produce three sp^2 hybrid orbitals of equivalent energy. In the formation of BF_3 molecule, the half –filled p orbital of each fluorine atom overlaps with each of the half-filled sp^2 hybrid orbitals of the boron atom.

In BF_3 three bonds are equivalent and are 120° apart. Therefore, the shape of BF_3 is triangular and planar.

NO_3^- ion : By X-ray studies, it has been shown that this ion has planar triangular shape. This reveals that the central nitrogen atom in this ion undergoes sp^2 hybridisation. But this ion is carrying a unit negative charge. Therefore, one of the three oxygen atoms (O_c) would be assumed to be having an extra electron.

Suppose three oxygen atoms are represented as O_a, O_b and O_c. The valence shell configurations of N, O_a, O_b and O_c are as follows :

$$N \longrightarrow 2s^2, 2p_x^1, 2p_y^1, 2p_z^1 \ \ (lp = 1)$$

$$O_a \longrightarrow 2s^2, 2p_x^2, 2p_y^2, 2p_z^2 \ \ (lp'\,s = 2)$$

$$O_b \longrightarrow 2s^2, 2p_x^2, 2p_y^1, 2p_z^1 \ \ (lp'\,s = 2)$$

$$O_c \longrightarrow 2s^2, 2p_x^2, 2p_y^1, 2p_z^1 \text{ (lp' s = 3)}$$

(Having an extra electron).

The 2s, $2p_x$ and $2p_z$ atomic orbitals of nitrogen undergo hybridisation to form three sp^2 hybrid orbitals. The half-filled $2p_y$ orbital does not take part in hybridisation. Thus, the electronic configuration of nitrogen atom becomes as follows :

$$N \rightarrow sp^2)^3, (sp^2)^1, (sp^2)^1, 2p_y{}^1 \text{ (lp = 1)}$$

Hybrid orbital having an electron pair forms a coordinate bond with O_a atom. The overlapping of half-filled hybrid orbital with half-filled 2p orbital of O_c – atom gives a covalent σ bond. Also, the overlapping of the remaining half-filled 2p-orbital of O_b–atom gives another covalent s bond. Half-filled 2p-orbital of N atom, which is not participating in sp^2 hybridisation, undergoes overlapping laterally with half-filled 2p orbital to form a pπ = pπ bond. Hence the structure of NO_3^- ion may be represented as follows :

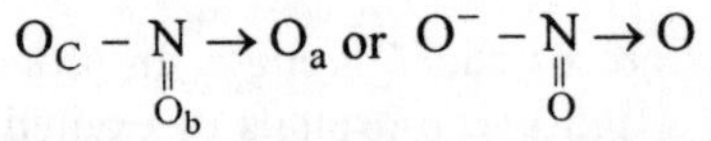

Double bond is made up of one σ and one π bond. This double bond does not alter planar trigonal and symmetrical shape of NO_3^- ion in which O–N–O angle is 120° Fig 4.15. In this all the three s bonds are sp^2 – p σ bonds.

Hybridisation of sp³ type : *The mixing of three p and one s orbitals to form four orbitals of equal energy is known as sp³ hybridisation.*

Characteristics : Following are the main characteristics of sp^3 hybrid orbitals.

(i) The four sp^3 hybrids are directed towards the four concerns of a regular tetrahedron whose centre is occupied by an atom that has undergone hybridisation.

(ii) The predicted relative overlapping power of each of the sp^2 hybrid is 2.00 whereas it is 1.93 and 1.99 for sp and sp^3 hybrid respectively. This shows that sp^2 hybrids are the strongest as compared to sp and sp^2.

(iii) The angle between hybrid orbitals should be 109° 28′.

Examples:

Typical examples involving *sp*^{3} hybridisation are.

(i) Shape of CH_4 : Carbon, the central atoms in CH_4, has only two unpaired electrons in the ground state. In the excited state, four unpaired electrons are available for bonding. The four bonds so formed are not of equal strength but actually they should be of equal strength. In order to explain, one s and three p orbitals undergo hybridisation and four equivalent sp^3 hybrid orbitals are formed. These four hybrid orbital can form four covalent bonds with hydrogen atoms to give a molecule of methane.

According to Sidgwick : Powell theory, the repulsion between four orbitals is smallest if they point to the corners of a regular tetrahedron, and make a bond angle of 109° 28′. The carbon atom is at the centre of the tetrahedron and four hydrogen atoms are at its corners.

(ii) Shape of Ethane : Ethane is another example of a molecule which utilises *sp*^{3} hybridisation. One of thc *sp*^{3} orbitals of one carbon atom overlaps with one of *sp*^{3} orbitals of the other to establish a C–C σ bond. The other *sp*^{3} orbitals of each carbon atom overlap with s-orbitals of hydrogen from C–H σ bonds.

(iii) Shape of NH_3 : In a molecule of ammonia, nitrogen is the central atom and has three unpaired electrons in its ground state sufficient to form three bonds.

Three unpaired electrons in the 2p orbitals are sufficient to form three bonds with three hydrogen atoms resulting ammonia molecule without involving any excitation of the atom. Thus, the molecule having three orbitals would have been triangular in structure and also symmetrical and non-polar.

But actually ammonia molecule is asymmetrical with a high dipole moment and is pyramidal in structure. In order to explain, it is postulated that the s-orbital (containing a lone pair of electrons) hybridises with three orbitals (each containing a bond pair of electrons) to give rise to four *sp*^{3} hybrid orbitals of equivalent energy.

According to the Sidgwick-Powell theory, the distribution of these four orbitals should be tetrahedral so that the force of repulsion between them is minimum.

The force of repulsion between a lone pair and a bond pair is greater than the force of repulsion between two bond pairs of electrons, therefore, the molecule gets a little distortion and the bond angle H–N–H decreases from 109.7° to 107°.

Thus, the shape of ammonia molecule is pyramidal having nitrogen atom at the centre, three hydrogen atoms at the base of the pyramid and the lone pair of electron at the apex of the pyramid.

(iii) Formation of water molecule : In a molecule of water, oxygen atom (centre atom) has two unpaired electrons which are sufficient to form two bonds with hydrogen atoms, without involving any excitation of the oxygen atom. Like BeF_2, H_2O molecule involving only two orbitals should be linear and non-polar.

According to some experimental evidences, water molecule has a bent structure and is highly polar. This latter structure of H_2O molecule can be explained on the basis of hybridisation. Two lone pairs of electrons, and the other two containing bond pairs undergo hybridisation to give four sp^3 hybrid orbitals of equivalent energy.

According to Sidgwick-Powell theory, these hybrid orbitals should result in a tetrahedral structure having two corners occupied by hydrogen atoms and two occupied by lone pairs of electrons as depicted.

	1s	2s	2p
Oxygen atom (ground state)	↑↓	↑↓	↑ ↑ ↑
Oxygen atom, having gained	↑↓	↑↓	↑↓ ↑↓ ↑↓

two electrons form hydrogen atoms in H_2O.

Here the two lone pairs distort the bond angle more than it was distorted by the one lone pair of electrons in ammonia ; the bond angle HOH in water in 104° 27′

sp^3d hybridisation : *The mixing of one s, three p-and one d-orbitals to form five hybrids sp^3d is known as sp3d hybridisation.* All these five hybrid orbitals are not of the same type but they can be divided into two non-equivalent sets.

(i) The first set consists of three equivalent hybrid orbitals which are oriented in a plane towards the corners of a triangle. This set is known as *equatorial set of orbitals* in which the hybrid orbitals are inclined at 120° to each other. This set is supposed to be formed by the hybridisation of one s, one $2p_x$ and one $2p_y$ orbitals forming three sp^2 hybrid orbitals.

(ii) The second set is known as *axial set* which consist of two equivalent hybrids, obtained by mixing one p_z and *one* $d_z{}^2$ orbitals. The hybrids of this set make an angle of 180° to each other but with the hybrids situated in the triangular plane they make an angle of 90°.

The electronic configuration of phosphorus is $1s^2\ 2s^2\ 2p^6\ 3s^2\ 3p^3$ and may be shown above. The phosphorus in its ground state has only three unpaired electrons. In order to explain the formation of PCl_5, five unpaired electrons are required. This can be achieved by the unpairing of electrons in the 3s orbital by the excitation of the atom. These five electrons (one s, three p and one d) undergo hybridisation to five hybrid orbitals. These five hybrid orbitals combine with five chlorine atoms to give PCl_5.

The structure is not symmetrical one and hence unstable. Thus PCl_5 is highly reactive and split into PCl_4^+ *and* PCl_6^- *ions in this solid state which have tetrahedral and octahedral structure respectively.*

Hybridisation of sp^3d^2 type : *The mixing of one 5, three p and two d orbitals to form 6 orbitals of equal energy is known as sp^3d^2 hybridisation. All the six hybrid orbitals in energy.*

sp^3d^3 hybridisation : The mixing of ones, three p and d orbitals to form seven orbitals of equal energy is known as sp^3d^3 hybridisation.

The sp^3d^3 hybrid orbitals are directed towards the corners of a pentagonal-bipyramid. These are not equivalent because five of them are directed towards the vertices of a regular pentagon whereas the remaining two are oriented perpendicular (90°) to the plane having the first set of five hybrid orbitals. Iodine heptafluoride is perhaps the only example of seven orbitals and sp^3d^3 hybridisation. The shape expected is pentagonal bipyramid.

Axial I–F distances have not been found to be same as equatorial I–F distances. This confirms the non-equivalent nature of sp^3d^3 hybrid or orbitals.

dsp^2 hybridisation : Combination of one d, one s and two p orbitals gives four dsp^2 equivalent hybrid orbitals. These are directed towards the corners of a square in xy-plane. In other words it means that these hybrids point away from the origin in the ± x and ± y directions of x- and y- axes.

dsp^3 hybridisation : Combination of one d, one s and three p orbitals gives five non-equivalent hybrid orbitals. These are directed towards the apices of a square pyramid. Out of the five hybrid orbitals, four are forming the basal set of equivalent orbitals whereas the remaining fifth is forming the axial orbital.

The number of outer orbitals and commonly occurring shapes and types of hybridisation are summarised.

Table 1

Number of outer orbitals	*Shape*	*Hybridisation*
2	Linear	sp
3	Plane Triangular	sp^2
4	Tetrahedral	sp^3
5	Trigonal bipyramidal	sp^3d
6	Octahedral	sp^3d^2
7	Pentagonal bipyramidal	sp^3d^3

OCTAHEDRAL COMPLEXES

The coordination number of the central metal atom or ion is six and hence these complex ions have octahedral geometry. This octahedral geometry arises due to d^2sp^3 or sp^3d^2 hybridisation of the central metal atom or ion. What type of hybridisation (*i.e.*, whether sp^3d^2 or d^2sp^3) will occur depends on the number of unpaired or paired electrons present in the complex ion. d^2sp^3 or sp^3d^2 hybridisation is also called octahedral hybridisation

Octahedral complexes in which the central atom is d^2sp^3 hybridised are called *inner-orbital octahedral complexes* while the octahedral complexes in which the central atom is d^2sp^3 hybridised are called outer-orbital octahedral complexes. Now let us see how d^2sp^3/sp^3d^2 hybridisation takes place in octahedral complexes.

d^2sp^3 HYBRIDISATION IN LINNER ORBITAL OCTAHEDRAL COMPLEXES

This type of hybridisation takes place in those octahedral complexes which contain strong ligands. On the basis of the orientation of the lobes of d-orbitals in space, these orbitals have been classified into two sets viz. t_{2g} or d_ε and e_g or d_r. t_{2g} set consists of d_{xy}, d_{yz} and d_{zx} orbitals while e_g set has df and $d^2{}_2$–y^2 orbitals. In the formation of six d^2sp^3 hybrid orbitals, two (n–1) d-orbitals of e_g set [*i.e.*, (n–l) $d^2{}_2$ and (n–1) $d_x{}^2$–y^2 orbitals], *one ns* and *three* np (npy npy and np,) orbitals combine together and form six d^2sp^3 hybrid orbitals.

Thus, we see that the two d-orbitals used in d^2sp^3 hybridisation are from penultimate shell [*i.e.* (n –l)th shell] while s and three p-orbitals are from ultimate shell (*i.e.*, nth shell). This discussion shows that in case of octahedral complex ions of 3d transition series elements, two d-orbitals used in d^2sp^3 hybridisation are $3d_z{}^2$ and $3d_{x^2-y^2}$ orbitals ($_g$ set of orbitals) while s- and p-orbitals are 4s and 4p orbitals. Thus d^2sp^3 hybridisation taking place in such complexes can be represented as $3d_{x^2-y^2}.3d_z{}^2.4p_x\ 4p_y\ 4p_z(d^2sp^3)$.

Since two d-orbitals used in d^2sp^3 hybridisation belong to inner shell [*i.e.*, (n – l)th shell], the octahedral complex compounds resulted from d^2sp^3 hybridisation are called inner orbital octahedral complexes.

Since these complexes have comparatively lesser number of unpaired electrons than the outer-orbital octahedral complexes (see later on), these complexes are also called *low spin* or *spin paired octahedral complexes*. It is due to the presence of strong ligands in inner-orbital octahedral complexes of 3d transition series that the electrons present in $3d_z{}^2$ and $3d_{x^2-y^2}$ orbitals (e_g set) are forced to occupy $3d_{xy}$, $3d_{yz}$ and $3d_{xz}$ orbitals (t_{2g} set) and thus 3d orbitals of e_g set become vacant and hence can be used in $3d_{x^2-y^2}\ .\ 3d_z{}^2.\ 4s.\ 4p_x\ 4p_y\ 4p_z\ (d^2sp^3)$ hybridisation.

Now let us discuss the structure of some octahedral complex ions of 3d transition series elements which result from d^2sp^3 hybridisation. In order to understand the concept of hybridisation, consider the case of carbon. Its atomic number is 6 and its electronic configuration is $1s^2\ 2s^2\ 2p_x{}^1\ 2p_y{}^1$.

	1s	2s	2p
Carbon (ground state)	↑↓	↑↓	↑ ↑ ↑
Carbon (excited state)	↑	↑ ↑ ↑	

In the carbon ground state, has two unpaired electrons and therefore it is bivalent. But in almost all its compounds, it shows a tetravalency.

In order to explain the tetravalency of carbon it is assumed that one of the 2*s* electrons becomes unpaired and gets excited to the vacant $2p_z$, orbital resulting in the configuration, $1s^2\, 2s^1\, 2p_x{}^1\, 2p_y{}^1, 2p_z{}^1$. Now carbon contains four unpaired electrons and, therefore, it is capable of forming four bonds.

If these four electrons now form four bonds, three bonds (involving the electrons $2p_x$ $2p_y$ and $2p_z$) would be of one type and at right angle to each other and fourth bond (involving the electron of 2s orbital) would be different with new directional properties.

But it is known with certainty that four bonds formed by the carbon atom are equivalent. In order to get four equivalent bonds, it is assumed that one of 2s electrons becomes unpaired, gets excited to the $2p_z$ orbital and then four orbitals (2s, $2p_x$, $2p_y$ and $2p_z$) get mixed up and finally redistribution of energy takes place between them, resulting in the formation of four equivalent hybrids.

This phenomenon of mixing of the atomic orbitals and the formation of new orbitals of equal energy is known as hybridisation and the new orbitals formed are known as hybrid orbitals. The bonds formed by such orbitals are called *hybrid bonds.* The compounds which are formed from these bonds are known as hybrids.

Conditions for Hybridisation

Following are the rules for hybridisation.

Rule I : *Only orbitals of similar energies belonging to the same atom or ion can undergo Hybridisation.*

Rule II : *Number of hybrid orbitals produced is equal to the number of orbitals mixed. Hybrid bonds are stronger than the single non-hybridised bonds of comparable energy.*

Rule III : *Most of the hybrid orbitals are similar but they are not necessarily identical in shape. They differ from one another in orientation in space.*

Rule IV : For equivalent hybrids, the orientation in space is determined by :

(i) *The number of orbitals mixed and consequently the number of hybrids obtained, and*

(ii) which of x, y and z directions are preferred by the orbitals when pure.

Rule V : *Once an orbital has been used to build a hybrid it is then no longer available to hold electron in its pure form.*

Rule VI : *The type of hybridisation indicates the geometry of molecules. From the type of hybridisation one can tell about the bond angles and structure.*

1. sp Hybridisation : A linear combination of one 5 and one p-orbital yields two *sp* hybrid orbitals. The mixing of orbitals is known as sp-hybridisation. This is illustrated in Fig 2.

Table 2

Shape of molecule	*Hybrid type*	*Bond angle*	*Examples*
Linear	*sp*	180°	$BeCl_2$, $HgCl_2$
Planar	sp^2	120°	BF_3,BEl_3, CO_3^{2-} NO_3^-
Tetrahedral	sp^3	109° 18'	CH_4, NH_4^+ BH_4^-
Square planar	dsp^2	99^d	$[Ni(CN)_4]^{2-}$, $(PtCl_4)^{2-}$
Trigonal bipyramidal	dsp^3	90° 120°	PF_5 (g)
Octahedral	$sp^3\,d^2$	90°	$(CrF_6)^{3-}$, SF_6, $[Co(NH_3)_6]^{3+}$

PROPERTIES OF SP-HYBRID ORBITALS

(i) Two sp-hybrid orbitals [sp_a and sp_b] are Completely equivalent in energy.

(ii) Each sp hybrid orbital is stronger than the pure s– and pure p-orbital from which it is formed after hybridisation. Its predicated relative overlapping power is 1.93.

(iii) In sp-hybrid orbital, there is one large lobe and one small lobe. The larger lobe, which is much bigger than the p-orbital, brings about a higher degree of overlapping with the orbital of a reacting atom. Thus, the resultant bond is comparatively stronger.

(iv) The normalised wave functions, $\psi\,(sp_a)$ and for sp_b hybrid orbital are as follows :

$$\psi\,(sp_a) = \frac{1}{\sqrt{(2)}}\,[\psi\,(s) + \psi\,(p)]$$

$$\psi\ (sp_b) = \frac{1}{\sqrt{(2)}}\ [\psi\ (s) - \psi\ (p)]$$

From the above equations, it follows that the ratio of p-character to s-character is 1 : 1 in each of the two hybrid orbital wave functions and that both the hybrid orbital wave functions are orthogonal to each other.

Example of sp-hybridisation : An interesting example involving sp-hybridisation is beryllium fluoride, BeF_2.

LDQR THEORY

In 1961, Linnett proposed an interesting modification of Lewis-Langmuir octed theory. According to him,

(i) The eight valency electrons present around atoms of first short period in their molecules should not be treated as four pairs but should be treated as two groups of four electrons. Each group of four electrons is termed as *quartet.* As eight electrons are assumed to be made up of two quartets in this theory, it is named as *Linnett double quartet theory (LDQT).*

(ii) Linnett further proposed that a set of four electrons (quartet) be considered to have the same spin, □, another quartet to have the opposite spin, ↓. In a quartet, the four electrons will naturally tend to arrange themselves in a way that their mutual repulsion effects are minimised. This is possible in the tetrahedral configuration.

The second quartet will also take up the tetrahedral configuration. Thus, the eight electrons arrange themselves in two tetrahedra. The two tetrahedra take up relative positions that keep the inter-electronic repulsion at a minimum.

(iii) This theory also considers the resonance contributions. However, it works best when an equal number of single and double bonds are present (SO_2 or CO_2).

(iv) This theory reduces the emphasis on electron pairing.

The mathematical treatment of Linnett theory is yet in the formative stage. This theory is intermediate between the VBT and MOT.

Let us illustrate the Linnett theory by considering the following examples :

1. Nitrogen Molecule : There are ten valency electrons between two nitrogen atoms in their molecule. This molecule can be represented by the Lewis and Linnett structures as follows :

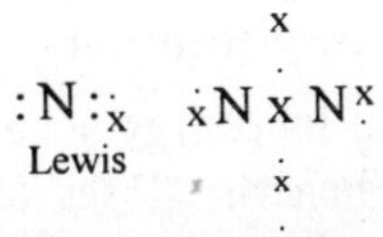

Lewis Linnett

According to both Lewis and Linnett the molecule is diamagnetic (paired spin) and the bond order is 3, *i.e.*, (6 – 0), Fig. 4.4 gives the LDQ representation where in there are two sets of five paired electrons.

The structure of carbon monoxide, acetylide ion (C_2^{2-}) and (CN^-) are similar.

Fluorine Molecule : In fluorine molecule, there are fourteen valency electrons. The Lewis structure of fluorine molecule is as follows :

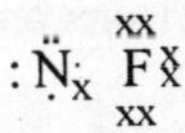

Lewis structure

In the LDQ structure, each fluorine atom has a quartet of electrons having the same spin, *i.e.*, four electrons denoted by x and four others by o. All the fourteen electrons are arranged in two spin sets of seven electrons. As the spins are equal in number and therefore cancel one another (7↑ and 7↑), the molecule is diamagnetic. Again, there are two electrons between the fluorine nuclei and hence bond order is one.

Oxygen Molecule : In this molecule there are sixteen electrons. According to the Lewis structure this molecule should be diamagnetic but actually it is paramagnetic.

:Ö::Ö:

Lewis structure

Linnett proposed the following structure :

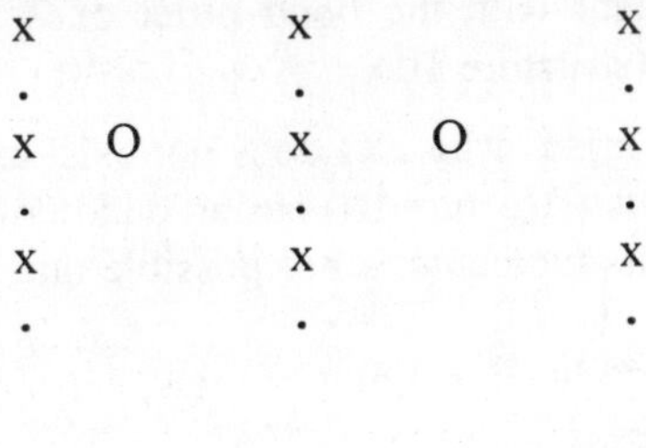

The above structure of Linnett actually represents two states for oxygen molecule. These are :

(a) One is the high energy state, $^{1}Sg^{+}$.. In this structure the two twin states are completely identical.

(b) The excited state. Its energy state lies 130 kJl mol^{-1} above the ground state. When this loses energy, it becomes stabilized giving die staggered arrangement ($^{1}\Delta g$) which is also diamagnetic because the electrons are still paired 6↑, 6↑.

(c) If the structure is formed from one spin set of five electrons and the second of seven (5↑, 7↑) the molecules is further stabilized. This is the $^{3}\Sigma g^{-}$ state and is more stable because its energy is at the ground state.

In this state the repulsion has been minimised, in spite of the total bonding electrons being four in number. In this structure there are two unpaired electrons and this shows that this molecule is paramagnetic.

Nitric Oxide Molecule : In this molecule there are 11 valency electrons having a set of 5 electrons of one spin (I) and 6 electrons of the opposite spin set (II). Each set has a tetrahedral orientation around the two atoms just as in O_2.

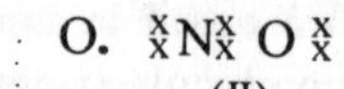

(III) (II)

The two structure (I) and (II) combine to yield structure (III).

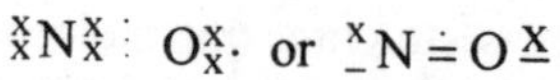

(III)

The structure (III) accounts for the presence of one unpaired electron in NO. The observed bond length in NO is intermediate between that expected for a double bond and a triple bond.

This is in agreement with the bond order of 2 1/2 in the LDQ representation of NO (structure III).

Now the question arises why NO does not exist as $(NO)_2$ because the nitric oxide molecule (structure III) has an odd number of electrons. But dimeric form of this molecule is not possible due to the following reasons.

(i) From the thermodynamic point of view, entropy is in favour of small molecules,

(ii) Dimerisation would take place if there is some gain in bonding. In the dimerisation of NO there is no gain of electrons because the bond order of 2 1/2 *i.e.* 1/2 (5–0) remains the same, in the dimer. If the dimer $(NO)_2$ were to be formed, one might expect it to dimerise through the pairing of odd electrons as shown below :

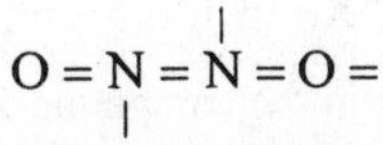

From the above discussion, it is valid that the dimer is unstable relative to the monomer.

GEOMETRY OF MOLECULES HAVING BOND PAIRS AND LONE PAIRS OF ELECTRONS

Due to the presence of lone pairs of electron in the valence shell, there occurs some distortion in the shape shape of the molecule.

Thus, the geometry of the molecule will not be the same as expected from typical cases of hybridisation. We shall now attempt to discuss some examples of molecules which have bond pairs as well as lone pairs of electrons.

(i) $SnCl_2$ Molecule : In stannous chloride, tin has electronic structure of outermost structure of outermost shell as : $5s^2$, $5p_x^{\ 1}$, $5p_y^{\ 1}$. In $SnCl_2$ two electrons are used for bond formation with two atoms of chlorine, while one electron pair is left unused as lone pair.

Thus, there are three electron pairs two bond pairs and one lone pair. The hybridisation is sp^2 and $SnCl_2$ is expected to be trigonal planar but actually it is bent or V-shaped because of the greater repulsion due to lone pair.

(ii) Structure of XeF_2 : Electric configuration of Xe = ... $5s^2 5p^6$ and the electronic configuration of Xe = $5s^2 5p^6 5s^1$ (2 unpaired electrons).

Thus, in xenon two unpaired electrons are availble which form covelent bonds with 2 fluorine atoms to give rise to the formation of XeF_2.

Orbital Representation of XeF_2

As XeF_2 involves sp^3d hybridisation, it means with three equatorial positions occupied by the lone electron pairs. The structure of XeF_2 has been investiaged by crystal structure study, Raman and infra-red spectra. All the study favors the trigonal bipyramid structure in both the vapour and crystalline state.

(iii) Structure of XeF_4 : Electronic configuration of Xe in the ground state = ... $5s^2 5p^6$.

Electronic configuration of Xe in the ground state = $5s^2 5p^6$

Electronic configuration of Xe in the compound XeF_4 = $5s^2 5p^4 5d^2$ structure of

Thus, in the formation of XeF_4, sp^3d^2 hybridisation is involved which gives rise to an octahedral shape with lone pairs projecting along the axis and fluorine atoms occupying the corners of a square plane having xenon at the centre.

(iv) Structure of ClF_3 : In this compound chlorine is the central atom and three fluorine atoms are arranged around it.

The electronic configuration of fluorine atom is

$$_9F = 1s^2\ 2s^2\ 2p_x^{\ 2}\ 2p_y^{\ 2}\ 2p_z^{\ 1}$$

The electronic configuration of chlorine atom is

$$_{17}Cl = 1s^2\ 2s^2\ 2s^6\ 2s^2\ 3p^5$$

and that of excited state of chlorine atom is

$$Cl\ (excited) = 1s^2\ 2s^2\ 2s^6\ 3s^2\ 3p^4\ 3p^5$$

Electronic configuration of Cl atom (excited state)

In the excited state of chlorine atom three unpaired electrons are available which form three bonds with three fluorine atoms involving sp^3d hybridisation. Therefore, ClF_3 may have trigonal bipyramid structures like I, II and III with two positions occupied by lone pairs.

It is not possible to predict which of the three possible arrangements I, II and III will be formed. From micro-wave date, the structure of ClF_3 is planar and T-shaped with bond angles almost 90°. This indicates structure (III). The slight distortion is due to the lone electron pair (IV).

(v) Structure of IF_5 : In this molecule, iodine is the central atom in the molecule.

The electronic configuration of fluorine atom is

$$_9F = 1s^2\ 2s^2\ 2p_x^{\ 2}\ 2p_y^{\ 2}\ 2p_z^{\ 1}$$

The electronic configuration of iodine atom is given as

$$_{53}I = 1s^2\ 2s^2\ 2p^6\ 3s^2\ 3p^6\ 3d^{10}\ 4s^2\ 4p^6\ 4d^{10}\ 5s^2\ 5p^5$$

and that of excited state is put as :

$$_{53}I\ (excited) = 1s^2\ 2s^2\ 2p^6\ 3s^2\ 3p^6\ 3d^{10}\ 4s^2\ 4p^6\ 4d^{10}\ 5s^2\ 5p^3\ 5d^2$$

ADDITIONAL EXAMPLES OF sp² HYBRIDISATION

1. Formation of carbonate ion, CO_3^{-2} **,**

$$C\ (excited) = 1s^2\ 2s^2\ 2p_x{}^1\ 2p_y{}^1\ 2p_z{}^1$$

$$O = 1s^2\ 2s^2\ 2p_x{}^1\ 2p_y{}^1\ 2p_z{}^1$$

$$O^- = 1s^2\ 2s^2\ 2p_x{}^2\ 2p_y{}^2\ 2p_z{}^1$$

In C–atm, one 2s and two 2p orbitals gets overlapped to yield three resultant equivalent sp^2 hybrid orbitals. The CO_3^{-2} ion has been carrying two until negative charges. Thus, two of the oxygen atoms may be regarded to be having one extra electron each.

(i) One of the sp^2 hybrid orbitals of carbon gets overlapped with 2p orbital of one oxygen atom to form $sp^2 - p$ sigma bond. Further unhybridised 2p orbital of the carbon atom gets overlapped with another 2p orbital to same oxygen atom to form p-p π-bond.

(ii) Second sp| hybrid orbital of carbon gets overlapped with half-filled 2p orbital of second oxygen atom (having an extra electron, *i.e.*, O^-) to form a sp^2-p sigma bond.

(iii) Third sp^2 hybrid orbital of carbon undergo overlapping with half-filled *2p* orbital of third oxygen atom possessing one extra electron, (O^-) to form a sp^2-*p* sigma bond.

Thus CO_3^{-2} ion is having *trigonal planar structure.*

Additional Examples of sp³ Hybridisation

1. Formation of sulphate ion (CO_4^{2-})

$$S\ (excited) = 1s^2\ 2s^2\ 2p^6\ 3s^1\ 3p_x{}^1\ 3p_y{}^1\ 3p_z{}^1\ 3s^{1\ +\ 1}$$

$$O = 1s^2\ 2s^2\ 2p_x{}^2\ 2p_y{}^2\ 3p_z1$$

$$O = 1s^2\ 2s^2\ 2p_x{}^2\ 2p_y{}^1\ 2p_z{}^1$$

One 3s- and three *3p* orbitals of S-atom undergo hybridisation to yield four sp^3 hybrid orbitals. Two *d* orbitals are unhybridised and are able to form π-bonds.

The sulphate ion is having two unit negative charges. Hence, two of the oxygen atoms may be having one extra electron each.

(i) Two of the four sp^3 hybrid orbitals of sulphur atom are able to overlap with 2p orbitals of two oxygen atoms (marked). (O-atoms) having sp^3-p-sigma bonds.

(ii) Remaining two sp^3 hybrid orbital of sulphur atom are able to overlap with 2p orbitals of two oxygen atoms (marked as O-atom) forming sp^3-p sigma bonds.

(iii) The two unhybridised orbitals of sulphur atom are able to overlap with 2py orbitals (still left unreacted) of two oxygen atoms marked as O-atom thereby forming pπ-dπ bonds.

Hence SO_4^{2-} ion is having *tetrahedral structure.*

2. Forming of perchlorate ion ClO_4^-

$$Cl\ (excited) = 1s^2\ 2s^2\ 2p^6\ 3s^1\ 3p_x{}^1\ 3p_y{}^1\ 3p_z{}^1\ 3d^{1+1+1}$$

$$O = 1s^2\ 2s^2\ 3p_x{}^2\ 3p_y{}^1\ 3p_z{}^1$$

$$O = 1s^2\ 2s^2\ 2p_x{}^2\ 2p_y{}^2\ 3p_z{}^1$$

One 3s and three 3p orbitals of Cl-atom undergo hybridisation to give four sp^3 hybrid orbitals. Three d orbitals are unhybridised and are able to form π-bonds.

The ClO_4^- ion is carrying unit negative charge. Thus one of the oxygen atoms may be regarded to be having one extra electron.

(i) One sp^3 hybrid orbital of Cl-atom undergoes overlapping with 2p orbital of oxygen atom marked (O-atom)- to field sp^3-p sigma bond.

(ii) Three sp^3 hybrid orbitals of Cl-atom undergo overlapping with 2p orbitals of three oxygen atoms to yield sp^3-p sigma bonds.

(iii) The three unhybridised d orbitals of Cl-atom overlap with 2p orbitals (still left unreacted) of three oxygen atoms marked as O-atom to form pπ-dπ bonds.

Thus ClO_4^- has tetrahedral structure.

3. Formation of chloride ion (ClO_3^-)

$$^{17}Cl\ (excited) = 1s^2\ 2s^2\ 2p^6\ 3s^2\ 2p_x{}^1\ 3p_y{}^1\ 3p_z{}^1\ 3d^{1+1}$$

$^{8}O = 1s^2\ 2s^2\ 2p_x{}^1\ 3p_y{}^1\ 3p_z{}^1$

$^{9}O^- = 1s^2\ 2s^2\ 2p_x{}^2\ 3p_y{}^2\ 2p_z{}^1$

One 3s and three 3p orbitals of Cl-atom intermix to give four sp3 hybrid orbitals. One sp^3 hybrid orbital contains a lone pair of electrons. Two d orbitals remain unhybridised and form p-bonds.

The ClO_3^- ion carries one unit negative charge. Thus, one of the oxygen atoms may be considered to have one extra electron.

(i) One sp^3 hybrid orbital of Cl-atom overlaps with 2p-orbital of oxygen atom marked as (O-atom) – to form sp^3-p sigma bond.

(ii) Two sp^3 hybrid orbitals of Cl-atom overlap with 2p orbitals of two oxygen atoms to form sp^3-p sigma bonds.

(iii) The two unhybridised d-orbitals of Cl-atom overlap with 2p orbitals (still left unreacted) of two oxygen atoms marked as O-atom, to form two pπ-dπ bonds.

The expected geometry of ClO_3^- ion is tetrahedral but due to the fact that lone pair bond pair repulsion is greater than bond pair bond pair repulsion, there is some distortion in the shape of molecule and assumes *pyramidal structure.*

Additional Examples of dsp² Hybridisation

1. Formation of tetracyanonickel (II) ion $[Ni\ (CN)_4]^{2-}$

The electronic configuration of nickel atom (At. No. 28) in ground state and Ni^{2+} ion , d_ϵ and d_γ represent d_{xx}, d_{yz}, d_{zx} and $d_x{}^2 - y^2$, $d_z{}^2$ orbitals respectively.

Electronic structure of nickel atom ground state Ni^{2+}

Ni^{2+} in strong octahedral ligand $[Ni(CN_4)]^{2-}$

As the cyanide ligands (: CN^-) approach the Ni^{2+} ion :

(i) The single electron in the $d_x{}^2 - y^2$ orbital gets repelled by four ligands while in dg orbital it gets repelled by only two ligands.

(ii) The d_γ electrons gets paired up in lower energy $d_z{}^2$ orbital while $d_x{}^2 - y^2$ orbital becomes vacant.

(iii) As $d_x{}^2 - y^2$ orbital remains vacant, the four CN^- ligands would readily easily approach along + x, – x + y and – y and – y directions.

(iv) As d_z^2 orbital is filled, the ligands approaching along + Z and –Z directions realise strong repulsion from filled d_z^2 orbital. Because of this, *square planar* [Ni $(CN)_4]^{2-}$ ion is formed. 'X' denotes pair of electrons denoted by the : CN^- ligands.

The formation of the square planar complex, $(PrCl_4)^{2-}$ ion (At No. Pt = 78, Cl, Cl = 17) could be similarly explained.

Additional Example of dsp^3 Hybridisation

Formation of tribromobis (triphenylphosphine) nickel (III); [Ni $(Et_3P)_2$ Br_3] complex

In this complex, the oxidation number of Ni has been + 3.

According to Hund's rule, the number of unpaired electrons in d-subshell of No^{3+} ion would be 3. To accommodate five electron pairs donated by two Et_3P; and three Br– ligands, there are required five vacant hybrid orbitals, dsp^3 (using inner 3d-orbital) or sp^3d (using outer 4d-orbital).

(a) dsp^3 hybridisation is attained by pairing the unpaired electrons in 3d subshell by the ligands. Because of this *trigonal bipyramidal structure with one unpaired electron is obtained.* It agrees with the experimental magnetic susceptibility results.

(b) sp^3d hybridisation is attained by pairing the unpaired electrons in 3d sub-shell has to be promoted to *high energy* vacant 4d orbital. The complex so formed would be energetically not favourable.

In this complex, it is expected two triethyl phosphine (Et_3P) molecules could be bond to nickel atom slight differently than other three bromine atoms. It has been found that two bonds called *axial three bonds* and *equatorial bond*).

ADDITIONAL EXAMPLE OF d^2sp^3 HYBRIDISATION

Hexacyanomanganese (II) ion [Mn $(CN)_6]^{4-}$

In this complex the oxidation state of manganese in [Mn $(CN)_6]^{4-}$ = + 2.

$[Mn(CN)_6]^{4-}$

Mn

Mn^{2+}

$[Mn(CN)_6]^{4-}$

According to Hund's rule, the number of unpaired electrons in d-subshell of Mn^{2+} ion would be 5.

So as to accommodate six electron pairs donated by six : CN^- ligands, six vacant hybrid orbitals, $d^3 sp^3$ (using inner 3*d*- orbital or $sp^3 d^2$ (using outer 4d-orbital) are available.

FORMAL CHARGE

It may be defined as *the charge that an atom (in a molecule) would appear to be having when all of the atoms are having the same electronegativity".*

The formal charge may be corresponding rather closely to real ionic charge. For example, in the ammonium ion (NH_4^+), the formal charge on each H–atom has been zero and that of N–atom has been +1, which corresponds to real ionic charge.

The formal charge may not be corresponding to real ionic charge. For instance, in the N_2O molecule, $(N_t = N_c = O)$, the formal charges on N_l N_c and O have been –1, + 1 and zero respectively. These charges are not equal to real ionic charge (zero) on the molecule.

Method to calculate formal charge on an atom. So as to find formal charge on an atom in a molecule or ion we have to follow the following procedure:

(i) Firstly electron dot structure (Lewis structure) of molecule or ion is first of all written.

(ii) All electrons are regarded to share equally in the molecule.

(iii) Each atom is regarded to be having one-half of the electrons it shares with neighbouring atoms.

(iv) Then, formal charge is calculated from the relation.

$$q_f = N_A - n_M$$

$$= n_A - n_L p - 1/2\ n_B p$$

where q_f refers to formal charge

n_A refers to number of electrons in the valence shell in the free atom.

n_M refers to number of electrons belonging to the molecule.

N_LP refers to the number of electrons in unshared pairs or lone pairs.

n_Bp refers to the number of electrons in the bonding pairs.

Example : *Calculate the formal charges on nitrogen and hydrogen atoms in ammonium ion* NH_4^+.

Solution : Valence electrons in N atom = 5

Valence electrons in H atom = 1

The electron dot representation of NH_4^+ *ion would be as follows* :

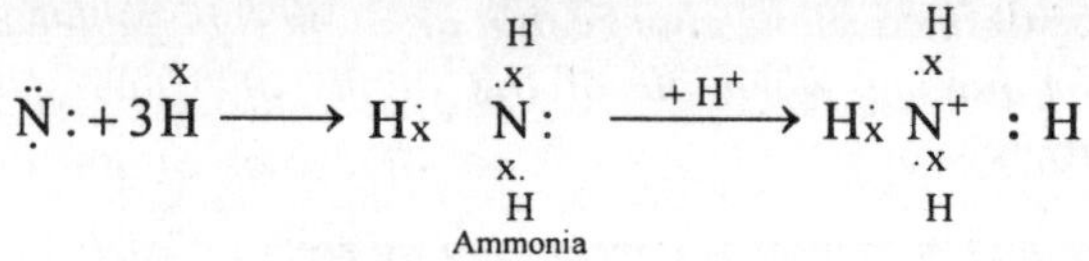

or

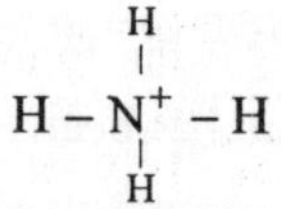

Ammonium ion

The formal charges can be calculated by using the relation

$$q_N = n_A - n_M = n_A - n_{Lp} - 1/2\ n_{Bp}, \text{ we have}$$
$$= 5 - 4 = 5 - 0 - 1/2\ (8) = +1 \text{ Ans.}$$
$$q_H = 1 - 1 = 1 - 0\ 1/2\ (2) = 0 \text{ Ans.}$$

ADDITIONAL EXAMPLES OF sp-HYBRIDISATION

1. Formation of carbon dioxide, CO_2 molecule

$$C \text{ (Ground state)} = 1s^2\ 2s^2\ 2p_x^{\ 1}\ 2p_y^{\ 1}\ 2p_z^{\ 0}$$
$$C \text{ (Excited state)} = 1s^2\ 2s^1\ 2p_x^{\ 1}\ 2p_y^{\ 1}\ 2p_z^{\ 1}$$
$$O = 1s^2\ 2s^2\ 2p_x^{\ 2}\ 2p_y^{\ 1}\ 2p_z^{\ 1}$$

In carbon atom (excited state), one 2s and one 2p orbitals undergoes overlapping to give two resultant equivalent sp hybrid orbitals. On the other hand p_y and p_z orbitals are unhybridised :

(i) Half-filled $2p_y$ orbitals of two oxygen atoms undergo overlapping with each sp–p hybrid orbital of carbon atom to yield two sp–p *sigma bonds*.

(ii) Half-filled $2p_z$ orbitals of two oxygen atoms undergo overlapping with unhybridised $2p_z$ orbitals of carbon atom to form two *pi*

(p) bonds. Because of this CO_2 is having *linear structure.*

The pictorial representation of CO_2 molecule is depicted in vig. 43.

2. Formation of acetylene (C_2H_2) molecule

$$C\ (\text{Excited}) = 1s^2\ 2s^1\ 2p_x^{\ 1}\ 2p_y^{\ 1}\ 2p_z^{\ 1}.$$

$$H = 1s^1$$

In carbon atom (excited state), one 2s and one 2p orbitals undergo hybridisation to yield to resultant equivalent. sp hybrid orbitals. *2py* and *2p,* orbitals are unhybridised.

One of the sp-hybrid orbitals undergo overlapping with *sp-hybrid* orbital of the other carbon atom to yield a (sp–sp) sigma bond. The remaining two *sp-hybrid* orbitals of two carbons undergo overlapping with the s orbitals of two hydrogen atoms to form two sigma bonds. The unhybridised p_y and p_z orbitals of two carbon atoms undergo overlapping laterally to form two π-bonds.

The combination of one (sp-sp) sigma bond and two (p-p) K-bonds is called the triple bond. Two π-electron clouds merge into each other to yield cylindrical n-electron cloud.

5

Theory of Coordination Compounds: Werner's and Pouling's

INTRODUCTION

The Werner (1893) put forward a theory, called Werner's theory of coordination compounds to explain the structure and properties of Co(III) and Pt(IV) ammines. These ammines are given below:

Co(III) ammines : $CoCl_3.6NH_3$ forange yellow), $CoCl_3.5NH_3.H_2O$ (pink), $CoCl_3.5NH)$ (violet), trans-$CoCl_3.4NH_3$ (green), cis-$CoCl_3.4NH_3$ (violet), $CoCl_3.3NH_3$ (blue-green).

Pt (TV) ammines : $PtCl_4.6NH_3$, $PtCl_4.5NH_3$, $PtCl_4.4NH_3$, $PtCl_4.3NH_3$, $PtCl_4.2NH_3$.

Different postulates of Wemer's coordination theory are given below:

1. **Two types of valency shown by the central metal atom in coordination compounds**

   In complex compounds, the central metal atom exhibits two types of valency viz. primary valenq (also called principal valency) and *secondary valency* (also called *auxiliary valency*). The metal atom always tends to satisfy both of its valencies.

2. **Characteristics of secondary valency**

(i) In modem terminology, the secondary valency of the metallic atom in a complex compound is equal to the coordination number of that metal.

(ii) Secondary valency of a metal is satisfied either by anions or by neutral molecules alone or by anions and neutral molecules both. For example.

(a) In $4KCN.Fe(CN)_2$ or $K_4[Fe(CN)_6]$, the secondary valency of Fe-atom is six and is satisfied by six CN^- ions (anions).

(b) In $CoCl_3.6NH_3$ or $[Co(NH_3)_6]Cl_3$, secondary valency of Co-atom is six and is satisfied by six NH_3 molecules (neutral molecules).

(c) In $CoCl_3.5NH_3$ or $[Co(NH_3)5Cl]Cl_2$, the secondary valency of Co-atom is six and is satisfied by five NH3 molecules (neutral molecules) and one CF ion (anion).

According to modem concept of complex compounds, the species satisfying the secondary valency are called ligands.

(iii) While writing the structure of a complex compound, the species satisfying the secondary valency and the metal are written inside the coordination sphere. For example in $CoCl_3.4NH_3$, since four NH_3 molecules and two CF ions satisfy the secondary valency of Co-atom, its structure is written as $[CoCl_2.4NH_3]Cl$ or $[Co(NH_3)4d_2]Cl$.

(iv) The secondary valencies have directional nature, since the species satisfying the secondary valency (*i.e.*, ligands) are directed towards the fixed positions in space.

(v) The number of species satisfying the secondary valency gives a definite geometry to the complex compound. For example, when the number of monodentate species (ligands) which satisfy the secondary valency of the metal in a complex compound is 2, 3, 4 or 6, the geometry of the complex compound would be linear, trigonal planar, tetrahedral or square planar and octahedral respectively In each of the four Co (III) ammines viz., $CoCl_3.6NH_3$, $CoCl_3.5NH_3$, $CoCl_3.4NH_3$ and $CoCl_3.3NH_3$, since the secondary valency of Co-atom is 6, each ammine has octahedral geometry.

(vi) The species satisfying the secondary valency cannot be obtained in the free state, when the aqueous solution of the complex compound undergoes ionisation. For example, when the aqueous solution of $CoCl_3.4NH_3$ or $[Co(NH_3)_4Cl_2Cl$ undergoes ionisation, four NH_3 molecules and two Cl^- ions which satisfy the secondary valency of Co-atom are not obtained in the free state as shown below:

$$[Co(NH_3)_4Cl_2]Cl \rightarrow [Co(NH_3)_4Cl_2]^+ + Cl^-$$

The above discussion can be concluded by saying that the species satisfying the secondary valency are not ionisable (non-ionisable) and hence cannot be precipitated.

(vii) The attachment of the species satisfying the secondary valency with the metal atom is shown by a solid line (–).

3. **Characteristics of primary valency :** (i) In modem terminology, the primary valency of the metallic atom in a complex'compound is equal to the oxidation state (or oxidation number) of that metal, *e.g.*, the primary valency of Co-atom in all the four Co(III) ammines is equal to +3.

(ii) Primary valency of a metal in a complex compound is always satisfied by onions, *e.g.*, the primary valency of Co-atom in each of the four ammines is equal to +3 and is satisfied by three Cl^- ions.

(iii) The anions satisfying the primary valency are written outside the coordination sphere while the anions which satisfy both the valencies (dual character) are written inside the coordination sphere. Thus, the species satisfyinig primary valency may be present inside and/or outside the coordination sphere.

(iv) The species satisfying both the valencies (*i.e.*, the species placed inside the coordination sphere) are directed towards specific directions in space and hence they have directional characteristics. The species which satisfy primary valency and are placed outside the coordination sphere have no directional characteristics.

(v) The species satisfying the primary valency do not give any geometry to the complex compound.

(vi) The species satisfying the primary valency can be obtained either completely or partially in their free state, when the complex compound undergoes ionisation in aqueous solution.

Examples (a) When the aqueous solution of $CoCl_3.6NH_3$ or $[Co(NH_3)_6]Cl_3$ undergoes ionisation, all the three Cl^- ions which satisfy primary valency of Co-atom are obtained in the free state.

$$[Co(NH_3)_6]Cl_3 \rightarrow [Co(NH_3)_6]^{3+} + 3Cl^-$$

(b) Although the primary valency of Co-atom in $CoCl_3..5NH_3$ or $[Co(NH_3)_5Cl]Cl_2$ is satisfied by three Cl^- ions, only two Cl^- ions are obtained in the free state, when the aqpeous solution of this ammine undergoes ionisation.

$$[Co(NH_3)_5Cl]Cl_2 \rightarrow [Co(NH_3)_5Cl]^{2+} + 2Cl^-$$

The above discussion shows that all or some of the species satisfying the primary valencies are ionisable and hence can be precipitated by means of a suitable precipitating agent. For example, in the above ammine two Cl^- ions can be precipitated as AgCl (white ppt.) by adding aqueous solution of $AgNO_3$.

$$[Co(NH_3)_5Cl]Cl_2 \rightarrow [Co(NH_3)_5Cl]^{2+} + 2Cl^-$$

$$2Ag^+ + 2Cl^- \rightarrow 2AgCl \text{ (white ppt.)}$$

In other words in understanding of co-ordination compounds and their properties began with the work of *Alfred Werner*. In 1899, Werner proposed an interpretation of co-ordination compounds which emphasized on the number and nature of groups attached to the metal ion. This theory replaced the older concepts of *Berzelius* (*1819*), *Graham* (*1837*), *Claus* (*1856*). *Blomstrand (1869) and Jorgensen* (*1878–1894*) and became a fundamental part of the electronic theory of valency formulated by *G. N. Lewis* (*1916*).

According to Werner,

(a) Metals possess two types of valencies, so called :

(i) Principal or ionisable valency, and

(ii) Secondary or non-ionisable valency.

(b) Primary valencies are those which a metal exercises in the formation of its simple salts. Thus primary valencies of Pt, Co, Cu and Ag in the formation of their simple salts, *e.g.*, $PtCl_4$, $CoCl_3$ and $CuSO_4$ and AgCl are 4, 3, 2 and 1 respectively.

(c) Secondary valencies are the valencies which a metal cation exercises towards a neutral molecule or an anion in the formation of complex ions. The secondary valency is also termed as the coordination number of the metal cation under consideration. Thus, co-ordination number gives the number of neutral molecules or onions which may be linked to the metal cation in the formation of its complexes. For example, cobalt (III) and platinum (IV) were recognised as having six secondary valencies (co-ordination number = 6), whereas copper (I) has four (co-ordination number = 4).

(d) Primary valencies are satisfied by negative ions whereas secondary valencies may be satisfied by either negative groups or neutral molecules.

(e) Secondary valencies are directed in space and hence such compounds are capable of exhibiting the phenomenon of isomerism.

On the basis of this theory, Werner assigned to $CoCl_3.6NH_3$ the structural formula given in Fig. 1.4. *by dashed lines.*

In this formula, size ammonia molecules are attached to the cobalt by secondary valencies where as three chlorine atoms are held to the cobalt by means of primary valencies. *Primary valencies are designated by solid lines and secondary valencies by dashed lines.*

EXPERIMENTAL EVIDENCE IN SUPPORT OF WERNER'S THEORY

A large number of complexes were studied by Werner and other workers, Werner used molar conductance values to determine the number of ions per molecule thus assigning groups to the first co-ordination sphere. Chemical methods like precipitation of Cl^- by $AgNO_3$ to determine the number of ionisable CF ions per molecule were also used.

Cobalt (III) ammine complexes : When an aqueous cobalt (II) chloride solution is treated with ammonia and then oxidised by air, the following six compounds are isolated :

(a) Luteo cobaltic chloride, $CoCl_3._6NH_3$: This is an orange yellow crystalline compounds. Its chief characteristics are :

(i) *Treatment of this compound in the solid state with sulphuric acid liberates all the chlorines as hydrogen chloride and leaves a sulphate. $Co_2(SO_4)_{3.}12NH_3$. Treatment of its solution with silver* *nitrate precipitates all the chlorines immediately. Thus, the bonding between cobalt and chlorine is purely ionic.*

(ii) *Treatment of solid with hydrochloric acid even at 100°C affects no removal of ammonia. Similarly, treatment of the solid with moist silver oxide gives a water soluble compound of composition $CoOOH.6NH_3$ which on addition of acids reforms the salts of type*

From the above experimental facts it follows that all the six ammonia molecules are unionisable and are held to the metal ion Co^{3+} in the first or inner coordination sphere whereas the ionisable Cl^- ions are in outer coordination sphere. From the above considerations, Werner formulated this complex as:

$$[Co(NH_3)_6]Cl_3$$

Conductance measurement indicates the presence of four ions, *i.e.*, one cobalt and three chloride ions (ammonia molecules are neutral). This is in accordance with the dissociation,

$$[Co(NH_3)_6\,Cl_3 \rightleftharpoons \underset{\text{Cmplex ion}}{[Co(NH_3)_6]^{3+}} + 3Cl^-$$

(b) Roseo cobaltic Chloride $CoCl_3.{}_5NH_3.H_2O$: This is a pink crystalline compound with the following characteristics :

(i) With $AgNO_3$, all the chlorines are precipitated as chloride from the solution of this complex indicating their ionizable character.

(ii) The water present in compound is firmly held at room temperature and is lost only at 100°C or above, showing its presence in the inner coordination sphere.

(iii) On treating with conc. H_2SO_4 it yields a compound $Co_2(SO_4)_3$ $10NH_3$. $5H_2O$ which on slight heating loses three water molecules, forming $Co_2(SO_4)_3$.$10NH_3$.$2H_2O$.

The above facts clearly indicate that water molecules and five ammonia molecules are firmly bound with cobalt whereas chlorines are loosely bound with cobalt. Thus, the formula of complex is

$$[Co(NH_3)_5.H_2O]Cl_3$$

Conductance measurements indicate the presence of four ions per molecule in very dilute solutions. This is in accord with the dissociation.

$$[Co(NH_3)_5.H_2O]Cl_3 \rightleftharpoons [Co(NH_3)5.H_2O]^{3+} + 3Cl^-$$

(c) Purpureo cobaltic chloride $CoCl_3.5NH_3$: It is a violet coloured compound with the following characteristics :

(i) When treated with conc. H_2SO_4, two chlorine atoms are lost as HCl, forming $Co(Cl)SO_4$.

(ii) Only two chlorines are precipitated immediately as AgCl on adding silver nitrate to its solution.

The above mentioned facts indicate the one chlorine atom is different from the rest two chlorine atoms. It means that only two of the three chlorines are ionisable as Cl^- ions.

On the basis of the above facts, Werner postulated the following formula : $[Co(NH_3)_5CL]Cl_2$

Conductance measurements indicate the presence of three ions per molecule in very dilute solutions. This is in accord with the dissociation.

$$[Co(NH_3)_5Cl]Cl_2 \rightleftharpoons [Co(NH_3)_5Cl]^{2+} + 2Cl^-$$

(d and e) Praseo-cobaltic chloride, violeo cobaltic chloride. $CoCl_3.4NH_3$: Two related compounds, one violet in colour (violeo-cobaltic chloride) and the other green (praseo-cobaltic chloride), have been prepared. Both have the same composition $CoCl_3.4NH_3$ but the chemical behaviour of these two compounds is quite different.

The conductances of solutions of these compounds show the presence of only two ions and from them only one chlorine is immediately precipitated by silver nitrate. Thus, from structural point of view their composition may be written as, $[Co(NH_3)_4Cl_2]Cl$

However, there were distinct differences in chemical behaviour between the two. Thus, although, only one third chlorine gets precipitated immediately on the addition of silver nitrate, second and third choleric atoms per molecule get also precipitated slowly and after long standing and no boiling.

Werner explained this by assuming that the chlorines of inner co-ordination sphere are slowly replaced by water molecules :

$$[Co(NH_3)_4Cl_2]^+ + H_2O \longrightarrow [Co(NH_3)_4H_2O.Cl]^{2+} + Cl^-$$

$$[Co(NH_3)_4H_2O.Cl]^{2+} + H_2O \longrightarrow [Co(NH_3)_4.2H_2O]^{3+} + Cl^-$$

However, the rate of precipitation of these chlorines in these compounds is different.

Werner explained the above facts by assuming that praseo-cobaltic chloride is trans- type whereas violeo-cobaltic chloride is cis- type.

(f) $CoCl_3.3NH_3$: This is a blue green compound having the following characteristics :

(i) *It does not give any precipitate immediately on addition of silver nitrate to its solution (absence of ionisable chlorine).*

(ii) *Its dilute solutions did not conduct electricity at all, showing the absence of any ions in solution.*

From the above facts Werner came to the important conclusion that all the groups are linked to the central metal ion by unionisable secondary valencies and gave the following formula : $[Co(NH_3)_3Cl_3]$

Werner applied his theory to the above ammonia compounds of cobalt (III) chloride. In all the compounds discussed above, the primary valencies are designated by solid lines and secondary valencies by dotted lines. In each case, satisfaction of six secondary valencies is essential. Werner's theory is successful in

(i) Predicting the correct number of ions in each case.

(ii) Firm association of chlorine with cobalt in the last three cases and

(iii) The existence of two forms of $CoCl_3.4NH_3$ is explained. It is due to difference in three dimensional arrangement of four ammonias and two chlorines about the central cobalt (III) ion.

In order to differentiate, materials held by secondary valencies are enclosed in square brackets when writing formulae whereas materials held by primary valencies outside the brackets. Thus the compounds just considered are written as.

$$[Co(NH_3)_6]Cl_3), [Co(NH_3)_5(H_2O)]\ Cl_3, [Co(NH_3)_5Cl]Cl_2$$
$$[Co(NH_3)_4Cl_2]\ Cl \text{ and } [Co(NH_3)_3Cl_3]$$

Criticism of Werner's Theory : One must appreciate the remarkable work done by Werner. He did this work 70 years ago and inspite of advances in theory and the increase in the number of coordination compounds, his postulates are as valid today as when they were presented over 70 years ago.

ELECTRONIC INTERPRETATION OF CO-ORDINATION

Sidgwick and Lowry gave an electronic interpretation of Werners theory.

According to Sidgwick, *Werner's primary valencies were regarded as amounting to electron transfer and the secondary or nonionic valencies to electron pair sharing.*

An important point is to be noted that *all neutral molecules or onions which were capable of being coordinated to central metal ion have atoms with at least one unshared lone pair of electrons in their valency shells.* Thus, the linkage of these coordinating groups to metal ion by secondary valencies arises due to the donation of the electron pairs to the central metal by these groups. *Bonds of this type are termed as co-ordinate bonds. These bonds have been indicated by arrows.*

Effective Atomic Number (EAN): Stability of the complex compounds is explained by the fact that central metal atom acquires the effective atomic number of the next inert gas. *The effective atomic number is defined as the number of electrons of the central metal in a complex including those gained by bonding.*

In most of the cases the effective atomic number is the same as the atomic number of the next inert gas. However, there are a few cases, where the effective atomic number is more or less by one or two units as compared to next inert gas. EAN of some metals in certain complexes are given Table 1. From the Table 2, it is evident that tendency to attain an inert gas configuration is a significant factor but not a necessary condition for complex formation.

Table 1 : Effective atomic number of some metals in complexes.

Atom	*Atomic number*	*Complex ion*	*Electrons lost in ion formation*	*Electrons gained by coordination*	*E.A.N number inert gas*	*Atomic of next*
Fe	26	$[Fe(CN)_6]^{4-}$	2	12	26–2+12=36	36 Kr
Co	27	$[Co(NH_3)_6]^{3+}$	3	12	27–3+12=36	36 Kr
Ni	28	$[Ni(CO)_4]$	0	8	28–0+8=36	36 Kr
Cu	29	$[Cu(CN)_4]^{3-}$-	1	8	29–1+8=36	36 Kr
Pd	46	$[Pd(NH_3)_6]^{4+}$	4	12	46–4+12=54	54 Kr
Cr	24	$[Cr(NH_3)_6]^{3+}$	3	12	24–3+12=33	36 Kr
Pd	46	$[PdCl_4]^{2-}$	2	8	46–2+8=54	54 Xe
Pt	78	$[Pt(NH_3)_4]^{2+}$	3	8	78–2+8=84	86 Rn

Some complexes of chromium and platinum are equally stable, yet the effective atomic number is short of inert gas configuration. For complex formation, it is essential to produce a symmetrical structure, *i.e.,* tetrahedral, square planar, octahedral, without any consideration of the number of electrons involved.

Objection. Sidgwick's electronic interpretation of co-ordination is not entirely satisfactory. Some objection are :

(i) The donation of electron pairs to a central cation would produce an improbable accumulation of negative charge on this ion. In the case of $[Co[NH_3)_6]^{3+}$ the donation of electron pairs from the ammonia molecules to the cobaltic ion would result an accumulation of negative charge on cobaltic ion.

In this way cobaltic ion would become negatively charged with respect to the ammonia molecules.

(ii) Another difficulty presented by this concept is the fact that the electron pair, available for donation in water, ammonia and many other neutral molecules, is $2s^2$ pair. The electrons have no bonding characteristics and to excite them to a higher energy level would require more energy than is usually available in bond formation.

Pauling's Theory

Pauling introduced in 1931, a new theory which was based on the revolutionary idea of hybridisation. His theory was able to account for and even predict the geometrical shapes of many complexes. According to him.

(i) Strong bonds are not formed by pure s-or pure p- or pure d-orbitals.

(ii) Strong bonds are generally formed by hybridised orbitals formed by a suitable combination of orbitals. For example, the combination of s-orbital and p orbitals gives four equivalent orbitals of equal energy known as hybrids. This type of combination is known as sp^3 hybridisation.

In complexes, other types of hybridisation like d^2sp^3, sp^3d^2 and dsp^2 are more common.

(iii) In transition mtal atom, d-orbitals are, partially filled up and hence it is the number of d-electrons in the central atom (forming complex) that determines the possible modes of hybridization and the shapes of complex.

Table 2

Bond Hybridisation	*Geometrical Shape*	*Typical Example*
d^2sp^3	Octahedral	$Fe(CN)_6^{3-}$
sp^3d^2	Octahedral	FeF_6^{3-}
sp^3	Tetrahedral	$Zn(NH_3)_4^{2+}$
dsp^2	Square planar	$Ni(CN)_4^{2-}$

(iv) In complex formation, Hund's rule of maximum multiplicity is strictly followed. However, in complex formation redistribution of electrons and complete pairing of electrons may take place. In this way some orbitals are vacated and made available for hybridisation.

Pauling's theory provides a suggestive explanation of the experimental facts and, in particular, of the magnetic susceptibilities of co-ordination compounds. Pauling's theory also differentiated between complex ions in which the bonding could be considered as ionic and those in which it was purely covalent.

Let us apply this theory to various examples;

Octahedral complexes : Octahedral complexes result from either sp^3d^2 or $d^2\,sp^3$ hybridisation as in the following examples.

(a) Ferricyanide ion, $[Fe(CNH)_6]3^-$: The structures of the Fe atom and the Fe^{3+} ion are given at (i) and (ii) of Fig. 1.68.

The five unpaired electrons in the Fe^{3+} give a calculated magnetic moment of 5.91 magneton. The magnetic moment for the $Fe(CN)_6^{3-}$ ion, however, is 2.3 suggesting, perhaps, one unpaired electron.

In order to explain the magnetic moment of $Fe(CN)_6^{3-}$, Pauling suggested that the unpaired *3d* electrons in the Fe^{3+} undergo pairing. The resulting complex would have one unpaired electron, and the d^2sp^3 orbitals are available for hybridisation.

The electrons, to occupy the d^2sp^3 hybrid orbitals, would be provided by the lone pairs from each of the six cyanide ion ligands. This means that the bonds between the ligands and the central ion are being coordinated linkages. Complexes like $|Fe(CN)_6]^{4-}$ are known as *inner-orbital complexes* because the *d-orbitals* involved are from a lower shell than the *s* and p-orbitals.

(b) Ferrocyanide ion $[Fe(CN)_6]^{4-}$: In the ferrocyanide ion there is one more electron than in the ferricyanide ion. This extra electron pairs with the unpaired 3d-electron in (iii), to give the structure shown in Fig. 1.8. The ferrocyanide ion has, therefore, no unpaired electrons and would be expected to be diamagnetic, this is found to be so. It is octahedral with d^2sp^3 hybridisation.

Other examples resembling structures like the ferri , or cyanide ions include $Mn(CN)_6^{3-}$, $Cr(NH_3)_6^{3+}$ and $Cr(CN)_6^{4-}$.

(c) Ferrifluoride ion $[FeF_6]^{3-}$: The $[FeF_6]^{3-}$ is octahedral like the ferri- and ferrocyanide ions. However, it has a magnetic moment of 6.0 which is in close agreement with the value of 5.9 – a value which would be expected from a structure with five unpaired electrons.

In order to explain this, it is assumed that no pairing of electrons in Fe^{3+} ions takes place and sp^3d^2 orbitals are used for hybridisation .

The FeF_6^{3-} is an example of an outer orbital complex because the d-orbitals involved in the hybridisation are from a higher shell than the s- and p-orbitals.

Other similar outer-orbital complexes are CoF_6^{3-} , $Fe(NH_3)_6^{2+}$: $Ni(NH_3)_6^{2+}$ etc.

Tetrahedral Complexes : Tetrahedral complexes result from sp^3 hybridisation. Complexes of Zn^{2+} are tetrahedral because they involve sp^3 hybrid bonds.

In the Zn^{2+} ion, all the $3d$ orbitals are fully occupied and therefore, they cannot take part in bond formation. However, sp^3 hybridisation takes place and four sp^3 hybrid orbitals are formed.

The electrons to occupy sp^3 hybrid orbitals would be provided by the lone pairs from each of the four ammonia molecules in $Zn(NH_3)_4^{2+}$. Thus, $Zn(NH_3)_4^{2+}$ will be tetrahedral. As there are no unpaired electrons, $Zn(NH_3)_4^{2+}$ will also be diamagnetic.

Other similar tetrahedral complexes are formed by Cd^{2+}, Hg^{2+}, Cu^+ and Ag^+ ions; all of which have the same structure as Zn^{2+}.

Square planar complexes : Square planar complexes result from dsp^2 hybridisation. This type of hybridisation involves mixing of one *d,* one *s-* and two 2*p* orbitals, resulting in the formation of four equivalent hybrid orbitals which are planar and mutually at right angles pointing towards the four corners of a square. The d–orbital involved is the dx^2-y^2 orbital in the shell lower than that of the *s-* and p-orbitals.

Square planar complexes tend to be formed when the central ions have only one d-orbital available in the lower shell.

(a) Nickelocyanide ion $Ni(CN)_4^{2-}$: The electronic arrangements for the Ni atom and the Ni^{2+} ion. If the two unpaired 3*d* electrons couple, the structure will be as given in (iii) so that dsp^{2-} hybrid bonds can be formed. All the electrons are paired so that the $Ni(CN)_4^{2-}$ ion is diamagnetic. Other similar examples are $PtCl_4^{2-}$, $PdCl_4^{2-}$, etc.

(b) Cuprammonium ion : $Cu(NH_3)_4^{2+}$: It has a coplanar square structure and is paramagnetic with a magnetic moment of 1.8. A square planar structure is obtained by dsp^2 hybridisation, keeping one electron unpaired.

OBJECTIONS TO THE SHAPE OF CUPRAMMONIUM ION

(a) The necessity to promote one electron from a 3*d* into a *4p* level is not satisfactory. The 4p electrons would be expected to be easily lost which would mean that the complex could be easily oxidised, but this is not so.

(b) Placement of the odd election in the 3*d* level as in (iv) would suggest sp^3 hybridisation with a resulting nonexistent tetrahedral structure.

Objections to Pauling's Theory : Pauling's theory explained the chemistry of co-ordination compounds satisfactorily but failed to explain some of the aspects :

(a) The peculiar behaviour of ions of divalent nickel, palladium, platinum and trivalent gold which apparently do not form the expected 5–co-ordination complexes.

(b) Pauling's theory proposed a form of hybridisation (dsp^2) suitable for square-planar structures but this does not explain why this structure is preferred to other possible stereochemical configurations, such as tetrahedral or 5–co-ordinated trigonal bipyramid which does, in fact, occur in iron carbonyl.

(c) Pauling's ideas offered a little or no explanation to the absorption spectra of complexes or of their relative stabilities or reactivities.

(d) It is well known that both the colour and magnetic moment of transition metal complexes are due to their possessing d-orbital electrons.

It means that there should be a quantitative connection between spectra and magnetic moment. Unfortunately this connection is not revealed by Pauling's theory.

(e) It cannot explain why certain complexes are more labile than the others.

Labile complexes are those in which one ligand can be easily displaced by another ligand. On the other hand, inert complexes are those in which displacement of ligands is slow.

(f) It does not take into consideration the splitting of d–energy levels.

(g) It does not account for the relative energies of different structures.

ACIDS AND BASIS

For instance acids turn blue litmus red while the bases change red litmus blue. Acids counter the effect of bases and vice versa. Some acids evolve hydrogen when reacting with some metals.

With the availability of large experimental data there arose a need to define acids and bases on a rational basis. The concepts of acids and bases developed one after another tend to make the definitions more and more broad based.

Here we will discuss the various concepts of acids and bases but with special emphasis on their modern concepts. Our recognition of some chemical substances as acids and certain others as bases was based on some experimental facts.

Acids and bases possess in a sense opposite characteristics. Acids and bases have been identified by their effects on some organic dyestuffs.

THE WATER IONS SYSTEM

According to Arrhenius theory of ionisation, all substances which give hydrogen ions when dissolved in water could be called acids while, those which ionise in water to form hydroxyl ions, coula be called bases.

Examples of Arrhenius acids are HCl, H_2SO_4, CH_3COOH, etc. Examples of Arrhenius bases are NaOH, NH_4OH, etc.

Acid-base equilibria could now be treated in a quantitative way and acid-base neutralisation could he considered to result from the union of the hydrogen ions with hydroxyl ions to form water molecules,

$$H^+ + OH^- \longrightarrow H_2O$$

When we write a general equation.

$$HA \longrightarrow H^+ + A^-$$

for an acid, suggesting the formation of a group, H^+, but this is not very correct as free protons could exist only when gaseous hydrogen atoms are ionised. In the presence of protophilic liquids such as water, unattached proton could not exist; in water a proton gets attached to a number of water molecules by co-ordinate linkages, releasing a very large amount of energy. The proton in its hydrated state is usually expressed by H_3O^+, even though the number of water molecules attached to one proton is more than one. The hydrated proton is called the hydroxonium or hydronium ion. On the other hand hydroxyl ions are also hydrated in water through hydrogen bonds and are represented by OH- formula only because of convenience.

Successes

(1) Arrhenius concept has been invaluable in elucidating the behaviour of aqueous solutions. The constant heat of neutralisation of a strong acid by a strong base is readily explained in the light of this concept.

(ii) This theory correlated the catalytic action of acids with the concentration of hydrogen ions.

(iii) This concept also lends to quantitative determinations of acid or base strengths front the determination of following equilibrium relation:

$$K = \frac{{}^{a}H^{+}\ {}^{a}H^{-}}{{}^{a}HB}$$

Limitations

(i) The Arrhenius concept lacks in generality.

(ii) The most serious limitation is that it deals with dissociation and acid-base reactions in aqueous medium only and does not mention the role of non-aqueous solvents in the dissociation of acids and bases into ions. According to this concept, HCl is regarded as an acid only when dissolved in H_2O and not in some other solvents like benzene, chloroform, etc. or when it exists in the gaseous form.

(iii) This concept restricts bases merely to hydroxides. Thus, even metal oxides would not be regarded as bases.

(iv) According to this concept, a neutralisation process is restricted to aqueous medium alone but many non-aqueous solvents like liquid NH_3 liquid SO_2, etc. are known to undergo comparable acid-base neutralisation reactions.

(v) It cannot explain the acidic behaviour of certain salts such as $AlCl_3$ in water.

(vi) This concept does not elaborate the fate of dissociated hydrogen' ions in solution because of their high hydrational energy (1075 kj/ mole).

(vii) A proton (H^+) is non-existent and is generally present as the hydrated H_3O^+ ion or H_9O_{4+} species.

(viii) A proton is also formed in other solvents like NH_3, ROH, etc.

(ix) There are nonprotonic and nonhydroxylic compounds which may yield H^+ or OH ions in water.

$$SO_3 + H_2O \rightarrow H_2SO_4 \rightleftarrows H^+ + HSO_4^-$$

$$CaO + H_2O \rightarrow Ca(OH)_2 \rightleftarrows Ca^{2+} + 2OH^-$$

PROTON-DONOR ACCEPTOR SYSTEM

Bronsted and Lowry in 1923 proposed a new concept to acids and bases which is independent of solvents. According to atom.

"An acid is any hydrogen contained substance (a molecule or a cation or an anion) whereas a base is any substance (a molecule or cation or an anion) that can accept a proton from any other substance." From the above definition, it follows that an acid is a proton donor and base is a proton acceptor. It also follows from this definition that it is not essential for bases to have OH^- groups as demanded by the Arrhenius theory.

A neutralisation process, therefore, involves a release of a proton by the acid, and the acceptance of that proton by a base. This concept makes acid-base reactions also to be known as protolytic reactions independent of solvent.

$$\underset{acid}{H:Cl} + \underset{base}{NH_3} \rightarrow Cl^- + NH_4^+$$

$$\underset{acid}{NH_4^+} + \underset{base}{OH^-} \rightarrow NH_3 + H_2O$$

Bronsted and Lowry acids and bases are of the following types :

Bronsted-Lowry Acids

Molecular : $HCl \rightarrow H^+ + Cl$

Cationic : $[A(H_2O)_6]^{3+} \rightarrow H^+ + [A(H_2O)_5OH]^{2+}$

Anionic : $H_2PO_4^- \rightarrow H^+ + HPO_4^{2-}$

Bronsted-Lowry Bases

Molecular : $Py + H^+ \rightarrow]PyH]^+$; where Py = Pyridine

Cationic : $[A(H_2O)_5OH]^{2+} + H^+ \rightarrow [A(H_2O)_6]^{3+}$

Anionic : $PO_4^{3-} + H^+ \rightarrow HPO_4^{2-}$

Substances like H_2O and HCO_3 ions who donate as well as accept protons.

$$H_2O \rightleftarrows H^+ + OH^- \text{ (donation)}$$

$$H_2O + H^+ \rightleftarrows H_3O^+ \text{ (acceptance)}$$

$$HCO_3^- \rightleftarrows H^+ + CO3 \text{ (donation)}$$

$$HCO_3^- + H^+ \rightleftarrows H_2CO_3 \text{ (acceptance)}$$

are classed as amphoteric substances

The Bronsted-Lowry concept is quite useful For example:

(i) This concept also takes into account protonic solvents like NH_3, H_2SO_4, etc.

$$NH_3 + NH_3 \rightleftarrows NH_4^+ + NH_2^-$$

$$H_2SO_4 + H_2SO_4 \rightleftarrows H_3SO_4 + HSO_4$$

(ii) All-proton transfer reactions may be treated as follows:

$$NH_4^+ + S^{2-} \rightarrow NH_3 + HS$$

Conjugate Acids and Bases

A study of the following reactions for processes occurring in water brings out another important feature of this concept about acids and bases:

$$CH_3COOH + OH^- \rightleftarrows HOH + CH_3COO^-$$
$$H_2O + H_2O \rightleftarrows H_3O^+ + OH^-$$
$$NH_4^+ + CO_3^{2-} \rightleftarrows HCO_3^- + NH_3$$
$$H_2O + CO_3^{2-} \rightleftarrows HCO_3^- + OH^-$$

From the above reactions, it is seen that when a molecule or an ion loses a proton, it is necessary to convert into a form which would accept a proton, to reform the original acid. The latter form is therefore, a base and it is called a linked or conjugate base to the initial acid.

Such pairs of substances which are formed from one another by the gain or loss of a proton are called conjugate acid-base pairs. Acetic acid and acetate ion, H_2O and OH^- ion and NH_4^+ ion and NH_3 , H_3O^+ ion and H_2O, HCO_4 ion and CO_3^{2-} ion are all examples of conjugate pairs of acids and bases. Just as for every weak acid we have a conjugate base, for every weak base we have a conjugate acid which is formed by the addition of a proton to the base.

The strong acids and bases are not covered by this statement for the lack of the corresponding proton combination.

Furthermore, a substance like HNO_3 behaving as an acid or a base depends on the solvent uses. In liquid NH_3, H_2O and even dry CH_3COOH, HNO_3 acts like an acid by donating protons, but in dry HF medium, it accepts protons, and behaves like a base.

Relative Strength of Acids and Bases

According to Bronsted concept, the strength of an acid is regarded as a measure of its tendency to donate proton. Thus, a strong acid is that one which has a strong tendency to donate a proton. Similarly, a strong bases is that one which has a strong tendency to accept a proton.

Generally, two methods are employed to compare the relative strength of given acids. These are as follows

(i) In the first method we compare the dissociation constants or ionisation constants of acids/bases to get idea about their strength. Higher the dissociation constant, greater the strength of acid/ base. For example, we are interested in comparing the acidic strength of CH_3COOH and HCN. The values of dissociation constant of CH_3COOH and HCN are 1.8×10^{-5} and 4.0×10^{-10} respectively.

$$CH_3COOH + H_2O \rightleftarrows H_3O+ + CH_3COO^-(K_a = 1.8 \times 10^{-5})$$

$$HCN + H_2O \rightleftarrows H_3O^+ + CN^- \quad (Ka = 4.0 \times 13^{-10})$$

As the value of K_a of CH_3COOH is greater than that of HCN, it means that CH_3COOH is a stronger acid than HCN and CN^- ion is a stronger base than CH_3COO^- ion

(ii) The second method is known as competitive protolysis method. In this method, one acid is added to the conjugate base of another acid, and, then, one has to determine the equilibrium concentrations experimentally.

For example, if $NaOC_2H_5$ is added to H_2O, it is exprimentally found that $C_2H_5O^-$ has good tendency to react completely with the H_2O to from C_2H_5OH and OH^- ion. It means that $C_2H_5O^-$ is a stronger base than OH^- and H_2O is a stronger acid than C_2H_5OH.

$$\underset{\text{Stronger acid}}{H_2O} + \underset{\text{Stronger base}}{C_2H_5O^-} \rightleftarrows \underset{\text{Weaker acid}}{C_2H_5OH} + \underset{\text{Weaker base}}{OH^-}$$

It is important to mention here that acidity increases with the increase of positive charge whereas the basicity increases with the increase of positive charge. Thus, $[Fe(H_2O)_6]^{3+}$ is a stronger acid than $[Fe(H_2O)_6]^{2+}$ ion and $[Ni(OH)_4]^{2-}$ is a stronger base than $[Ni(OH)_4]^-$ ion.

Successes

(i) This concept shows that acid base phenomena are dependent on the solvent.

(ii) The concept covers a large number and variety of substances under base, for example, hydroxide ion, amide ion, ethoxide ion, pyridine, alcohols, acetate ion, bisulphate ion and many others.

(iii) Bronsted-Lowry concept is useful in accounting for the hydrolysis of salt solution. A solution of $FeCl_3$ is acidic because the proton donor ability of hydrated ferric ion exceeds the proton acceptor ability of the chloride ion and measurable excess of $[Fe(H_2O)_x]^{3+}$, ion appears in the solutions.

$$FeC1_3 + {}_xH_2O \rightleftarrows [Fe(H_2O)_x]^{3+} + 3Cl^-$$

Limitations

This theory serves excellently in protonic solvents but fails in case of some obvious acid-base reaction. For example, the formation of H_2SO_4 by the reaction of SO_3 and H_2O, the acidic participant SO_3 in this case has no protons to give.

Levelling Solvents

The apparent strength of a protonic acid has been found to depend upon the solvent in which the acid is dissolved. When all the acids which are stronger than H_3O^+ (*e.g.*, $HClO_4$, H_2SO_4, HCl and HNO_3) are added to H_2O, they lose a proton to H_2O forming H_3O^+ ion.

All these acids have equal strength because all these acids have been levelled to the strength of H_3O^+ ion which remains in solution and is common to all such solutions. This phenomenon in which the strength of all the acids becomes equal to that of H_3O^+ ion is known as levelling effect of the solvent *i.e.*, water. Here water is known as a levelling solvent for all these acids.

Solvent	*Acid*	+ *Base*	→ *Salt*	+ *Solvent*
H_2O	H_3O^+ (HBr)	OH^- (KOH)	K^+, Br^-	$2H_2O$
C_2H_5OH	$C_2H_5OH_{2+}$ (HBr)	$O^-C_2H_5$ (KOC_2H_5)	K^+, Br^-	$2C_2H_5OH$

The solvents in which complete transfer takes place are known as levelling solvents. It means that the solvents in which the solute is 100% ionised are known as levelling solvents. For example, liquid NH3 acts as a levelling solvent for HF and HCI because both these are -100% ionised in liquid NH_3 to give-100% NH_4^+ ions.

In H_2O, HF undergoes partial ionisation while HCI and HBr undergo –100% ionisation. Thus, H_2O is a differentiating solvent for HF but for HCl and HBr it is a levelling solvent.

Other solvents may have their own levelling effects, too. Thus, benzoic acid is a very weak acid in aqueous solution.

But in liquid ammonia it acts as a very strong acid.

In general, it may be expected that acidic solvents (like acetic acid) will exert less levelling effect on acid strengths (and similarly-basic solvents on base strengths). Relative strengths of the common mineral acids have been obtained by conductace measurement in anhydrous acetic acid. The order found is

$$HClO_4 > HBr > H_2SO_4 > HCl > HNO_3$$

CONCEPT OF THE SOLVENT SYSTEM

Franklin extended the dissociation principles to non-aqueous solvents and realised that non-aqueous solvent molecules may also dissociate into two oppositely charged ions. According to him, if an acid releases H^+ in water to form H_3O^+ ions then an ammonium salt, NH_4Cl, by releasing NH_4^+ in liquid ammonia, must also be termed an acid.

If NaOH, by virtue of its releasing OH^- ion in aqueous medium, is termed a base in that medium, then $NaNH_2$ should also be termed a base in liquid NH_3 since it releases the amide, NH_2^- ion. The dissociation (or auto-ionisation) of non-aqueous solvent is responsible in such solvents.

$$NH_3 \rightleftarrows H^+ + NH_2^-$$

According to Franklin, an acid is a substance which by dissolution in the solvent forms the same cation as does the solvent itself due to autoionisation. A base is one that gives on dissociation in the solvent the same anion as does the solvent itself on its own ionisation.

Solvent	*Acid +*	*Base →*	*Salt +*	*Sovent*
H_2O	H_3O^+ (HBr)	OH^- (KOH)	K^+, Br^-	$2H_2O$
C_2H_5OH	$C_2H_5OH_2^+$ (HBr)	$O^-C_2H_5$ (KOC_2H_5)	K^+, Br^-	$2C_2H_5OH$

The solvent system concept to non-protonic solvent *e.g.*, SO_2, $COCl_2$ in case of acids and bases was applied by Cady and Elsey.

According to them, substances which increase the concentration of the cation characteristics of the solvent are acids, whereas substances which increase the concentration of the anion characteristics of the solvents are bases.

In N_2O_4 as a solvent, substances such as NOCl which yield NO^+ ions behave as acids, and substances such as $NaNO_3$ which yield NO_3^- ions behave as bases.

The auto-ionisation of some non-protonic solvents is given below:

Acid Base		Acid Base
$SO_2 + SO_2$	$\rightleftarrows$	$SO^+ + SO_3$
$COCl_2 + COCl_2$	$\rightleftarrows$	$.COClCOC_2^+ + Cl^-$
$N_2O_4 + N_2O_4$	$\rightleftarrows$	$2NO^- + 2NO_3^+$
$BrF_3 + BrF_3$	$\rightleftarrows$	$BrF_2 + BrF_4^-$

Just as with Arrhenius concept neutralisation is reaction between an acid and a base to produce a salt and the solvent. Neutralisation reactions in non-aqueous media are given below:

Acid Basc Salt Solvent

In liquid NH_3: $NH_4Cl + NaNH_2 \rightleftarrows NaCl + 2NH_3$

In liquid SO_2: $SOCl_2 + Na_2SO_3 \rightleftarrows 2NaCI + 2SO_2$

It is clear from the following reactions that there is complete analogy between solvolytic and amphoteric behaviours in aqueous and non-aqueous solvents.

Solvolytic Behaviour

$$\left[\begin{array}{ll} \text{In liq. } NH_3 & : AlCl_3 + NH_3 \rightarrow [Al(NH_2)]^{2+} + H^+ + 3Cl^- \\ \text{In } H_2O & : AlCl_3 + H_2O \rightarrow [Al(OH)^{2+} + H^+ + 3Cl^- \end{array}\right.$$

Amphoteric behaviour:

$$\left[\begin{array}{l} \text{In liq } NH_3: Zn(NH_2)_2 + 2NH_2^- \rightarrow [Zn(NH_2)_4]^{2-} \\ \quad \text{In } H_2O : Zn(OH)_2 + 2OH^- \rightarrow [Zn(OH)_4]^2 \end{array}\right.$$

Success

This general theory of solvent system concept possesses the obvious advantage of relating acid-base behaviour in aqueous and non-aqueous solvents, both protonic and nonprotonic.

Limitations

(i) It limits acid-base phenomena to solvent systems only.

(ii) The concept does not explain the neutralisation occurring without the absence of solvent.

(iii) It does not explain the acid-base processes which may occur in the absence of solvent.

(iv) In low dielectric constant solvents, the existence of ions is not attainable.

(v) Emphasis has been laid mainly on the chemical properties of the solvent. However, the physical properties have been completely ignored.

(vi) Too much stress has been laid mainly on the ionic reaction which actually even dielectric affects would be able to explain the reactions easily.

For instance, Gutman and his coworkers (1952-60) have explained reactions between the chloride ion acceptors ($FeCl_3$,$SnCl_4$, etc.) and chloride ion donors (Me_4NCl) in $OPCl_3$ as follows:

$$OPCl_3 \rightarrow OPCl_2^+ + Cl^- \text{ (autoionisation)}$$

$$Me_4NCl \rightarrow Me_4N^+ + Cl^- \text{ (base)}$$

$$OPCl_3 + FeCl_3 \rightarrow OPCl_2^+ + FeCl_4^- \text{ (acid)}$$

However, Meek and Drago (1962) showed that the reaction can be carried out equally well in triethyl phosphate medium without involving chloride ions.

It means that the dielectric constant of the medium is more important than the ionisation reaction of the solvent.

Thus the solvent system concept for acid and bases simply extends, but does not necessarily improve, the Arrthenius waterions system.

LUX-FLOOD CONCEPT

Lux proposed a theory to explain the acid-bases concept of oxide system. According to the concept, a base is any material which gives up oxide ions and an acid is any material which gains oxide ions. Some typical reactions are

$$\text{Base} \rightleftarrows \text{Acid} + {}_xO^{2-}$$

$$SO_4^{2-} \rightleftarrows SO_3 + O^{2-}$$

$$CaO \rightleftarrows Ca^{2+} + O^{2-}$$

Hence according to the concept, a bases is an oxide donor and acid is an oxide ion acceptor. This view is particularly useful to high temperature chemistry, as in the fields of ceramics and metallurgy.

For example, the following reactions

$$CaO + SiO_2 \rightarrow CaSiO_3$$

$$PbO + SO_3 \rightarrow PbSO_4$$

involving basic oxides (CaO,PbO) and acidic oxides (SO_3, SiO_2) are reacting to form salts.

Furthermore, according to this theory amphoteric substances are those which show both a tendency to take up or give up oxide ions depending upon the circumstances, *i.e.*,

$$Na_2O + \underset{\text{Acid}}{ZnO} \rightarrow 2Na^+ + ZnO^2$$

The oxide transfer picture due to Lux can be extended to any negative ion, *i.e.*, halide, sulphide, etc.

$$3NaF + AlF_3 \xrightarrow[\text{temperature}]{\text{high}} 3Na^+ + AlF_6^{2-}$$

The Lux-flood definition can be extended to include transfer of any anion *e.g.*, halide, sulphide, etc.

Limitations

The limitations of such a concept are obvious. This view can be reduced to one part of the more general theories of Lewis and Usanovich.

LADY-ELSEY CONCEPT

Bases on the characteristic cation or anion of the solvent which is produced by it with the solute, a more general definition of acids and bases applicable to all solvents was provided by Cady and Elsey. Consider, for example.

(i) $$H_2O + H_2O \rightarrow H_3O^+ + OH^-$$

(ii) $$NH_3 + NH_3 \rightarrow NH_4^+ + NH_2^-$$

(iii) $$HF + HF \rightarrow H_2F^+ + F^-$$

(iv) $$N_2O_4 + N_2O_4 \rightarrow 2NO^+ + 2NO_3^-$$

(v) $$CH_3COOH + CH_3COOH \rightarrow CH_2COOH_2^+ + CH_3COO^-$$

The characteristic cations furnished by water, ammonia, hydrogen fluoride and nitrogen tetroxide from (i), (ii), (iii) and (iv) above are respectively H_3O^+, NH_4^+, H_2F^+ and NO^+. The respective characteristic anions of these solvents are OH^-, NH_2^-, F^- and NO_3^-.

According to Cady-Elsey Concept

An acid is a substance which in dissolution in solvent gives the cations characteristic of the solvent and a base is a substance which yields anion characteristic of the solvent either by direct dissolution or by reaction with the solvent.

In the light of this concept, it is seen from (i), (ii), (iii) and (iv) above that:

All substances which furnish H_3O^+ ions in water solvent, NH_4^+ ions in ammonia solvent, H_2F^+ in hydrogen fluoride solvent, NO^+ in nitrogen tetroxide and acetic acid medium are termed as bases. For example,

$$NH_4X \rightleftarrows NH_4 + X^-$$

in ammonia is an acid and $MNH_2 \rightarrow M^+ \ NH_2^-$ in it is ao base. BF_3 in hydrogen fluoride reacts as $BF_3 + 2HF \rightleftarrows H_2F^+ + BF_4^-$ forming H_2F^+ ions. It is therefore an acid.

ELECTRON-DONOR-ACCEPTOR SYSTEM OR THE LEWIS CONCEPT

Gilbert N. Lewis proposed a broader concept of acids and bases that liberated acid-base phenomena from the proton; although Lewis first proposed his system in 1923, he did little to develop it until 1938, Lewis defined a base as a substance that has an unshared electron pair with which it can form a covalent bond with an atom, molecule, or ion.

An acid is a substance that can form a covalent bond by accepting an electron pair from a base.

An acid must have a vacant orbital into which an electron pair donated by a base can be accommodated.

An example of a Lewis acid-base reaction that is not treated as such by any other acid-base concept

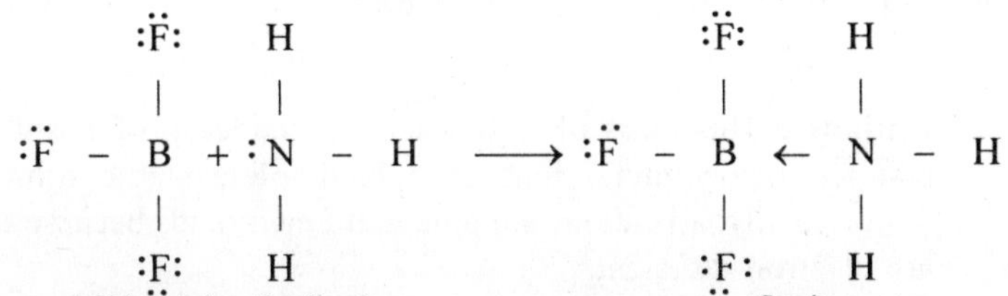

Acid (acceptor) (Electrophile) Base (donor) (Nucleophile) Complex

Lewis acid is often described as acceptor or electrophile, and the Lewis base is called a donor and nucleophile. On the basis of this concept, cations are acid-base complexes.

Classification of Lewis acids and bases : Based on the Lewis system are essentially the same as in Bronsted-Lowry system, since any molecule or ion which accepts protons does so because of the presence of an unshared pair of electrons. For example, OH , CF^-. NH_3, H^3O^+, CN^- etc., are bases on the Lewis as well on Bronsted-Lowry system.

The properties of an acid might be due to the availability of an empty orbital for the acceptance of a pair of electrons. Hence Lewis definition does not attribute acidity to any particular element but rather, to a unique electronic arrangement. According to this concept, acids are classified as

1. **Molecules possessing a central atom with less than an octet of electrons :** Lewis acids of this class are the electron deficient molecules such as B in BF_3, Al in $AlC1_3$.

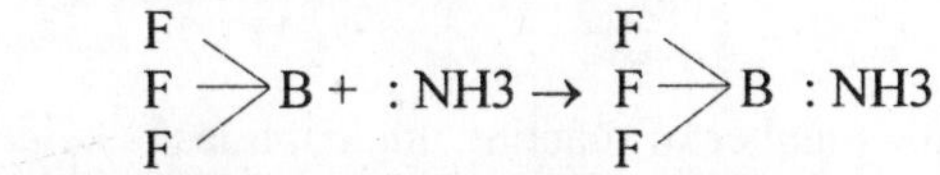

Acid (acceptor) (Electrophile) Base (donor) (Nucleophile) Complex

2. **Molecules in which the central atom has available 'd' orbitals and may acquire more than an octet of valence electrons:** Silicon tetrafluoride and tin (IV) chloride are typical examples:

$$\underset{\text{Lewis acid}}{SiF_4} + \underset{\text{Lewis base}}{2:\ddot{\underset{..}{F}}^-:} \rightarrow \underset{\text{Complex}}{SiF_6^{2-}}\ ;\ \underset{\text{Lewis acid}}{SnCl_4} + \underset{\text{Lewis base}}{2:\ddot{\underset{..}{Cl}}^-;} \rightarrow \underset{\text{Complex}}{SCl_6^{2-}}$$

Some other examples of this type are

PF_3, PF_5, $TeCl_4$, SF_4, SeF_4, TiX_4, GeX_4,

These halides tend to form adducts with a number of organic bases.

3. **Cations :** This class of Lewis acids is made up of positively charged heavy metal ions with incomplete stable orbitals. Theoretically, all cations are potential Lewis acids because they are electron deficient.

   However, this property is negligible for the alkalimetal cations such as Na^+, and K^+, and is weak for the alkaline earth cations such as Ca^{2+}, Sr^{2+} and Ba^{2+}. Of the alkaline earth group, only Mg^{2+} and Be^{2+} show an appreciable tendency to behave as Lewis acid. The reaction of M^{3+} type of ion with water molecules is an example of Lewis acid-base interaction.

$$\underset{\text{acid}}{M^{3+}} + \underset{\text{Base}}{6H_2O} \longrightarrow$$

(where M = Al, Cr, Fe)

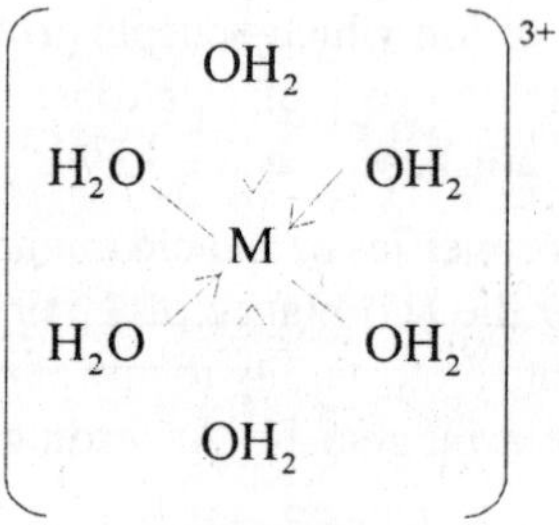

Complex

The reaction of Ag^+ with NH_3 is also typical

$$\underset{\text{Lewis acid}}{Ag^+} + 2:\underset{\text{Lewis base}}{NH_3} \rightarrow \underset{\text{Complex}}{[H_3N:\ Ag\ :NH_3]^+}$$

Usually, the number of reacting nucleophiles is twice the charge on the cation.

4. **Molecules with a multiple bond between atoms of dissimilar electronegativities :** An example of such a molecule is carbon dioxide, O = C = O. The oxygen atoms are more electronegative than the carbon atom. As a result, the electron density due to the it electrons is displaced away from the carbon atom, towards the oxygen atoms. The carbon atom is electron deficient and is apt to from a bond with a Lewis base such as OH^-

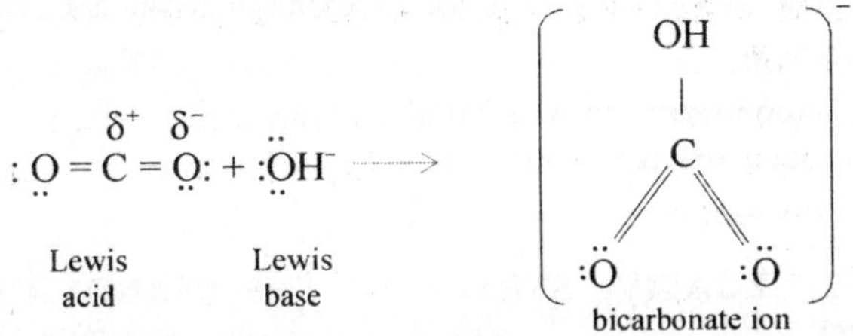

bicarbonate ion

SO_2 reacts like CO_2 towards OH^-,

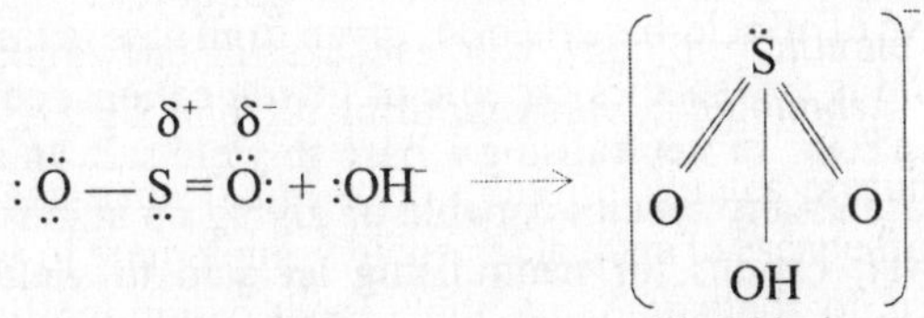

Bisulphite ion

5. **Elements with an electron sextet :** Oxygen and sulphur atoms contain six electrons in their valency shells and, therefore, act as Lewis acids. The oxidation of sulphite to SO_4^{2-} by oxygen, and to $S_2O_3^{2-}$ by sulphur involves acid-reactions.

$$:\ddot{\underset{\cdot\cdot}{O}}-SO_3^{2-} \rightarrow (O \leftarrow SO_3)^2$$

$$:\ddot{\underset{\cdot\cdot}{S}}-SO_3^{2-} \rightarrow (S \leftarrow SO_3)^2$$

Successes

(i) Lewis concept is a broader interpretation of acid- base behaviour.

(ii) This theory is not dependent on the presence of one particular element, upon any given combination of elements, upon the presence of ions, or upon the presence or absence of a solvent.

(iii) The Lewis approach is, however, of great value in case where protonic concept is not applicable.

(iv) Lewis theory is frequently employed to interpret reaction mechanisms.

Limitations

(i) In Lewis approach there is a lack of uniform scale of acid and base strength. The strength of acid and base in terms of Lewis concept is variable and dependent on the reaction considered.

In this respect it is inferior to Arrhenius concept or the Lowry concept.

(ii) The concept is so broad that it is unsuited to general chemistry, in which the explanation of acid-base behaviour is most important of an acid.

POSITIVE-NEGATIVE SYSTEM OF THE USANOVICH CONCEPT

Usanovich proposed definitions of acid and bases which are most comprehensive of all acid-base theories given from time to time. According to him, an acid is any species capable of giving cations, combining with anions or electrons, or neutralising a base to yield salt. In a similar way he defined base as any species capable of giving up anions or electrons, combining with cations, or neutralising an acid to yield a salt. The Usanivich theory includes all precious acid-base definitions and in this process it has become too general. It also includes oxidation-reduction reaction as a special class of acid-base reactions.

Usanovich has placed considerable emphasis on the central atom in the compound while accounting for the acid-base character of a substance. By co-ordination unsaturation he means the ability of an atom to increase its covalence.

The acid function is determined by the presence of co-ordinately unsaturated positive particles while basic function by the presence of similarly unsaturated negative particles. For instance, in SO_2 the central atom can be considered to be sulphur, and it is co-ordinately unsaturated. Thus, it is capable of accepting an anion such O^{2-}, thereby acting as an acid. In general acidity is promoted by highly charged positive particles (Na_2O is highly basic while Al_2O_3 is less basic) and basicity by highly charged negative particle.

Acid Base Relationships on the Basis of Usanovich Concept

Acid +	Base →	Salt	Justification
Cl_2 +	2Na →	2NaCI	Na loses an electron, Cl, gains, an electron
SO_2 +	Na_2O →	Na_2SO_4	Na_2O yields O^{2-}
Sb_2S_5 +	$3(NH_4)_2S$ →	$2(NH_4)3SbS_4$	$(NH_4)_2$ S gives up S^{2-}, Sb_2S_5 gains S^{2-}

Success

Usanovich concept is useful in classifying all examples of acidity and basicity. It is most general for all other definitions of acid-base so far given.

Limitations

(i) Usanovich concept is of great value for specific problems, but for general considerations, it is undoubtedly too broad to be of wide applicability.

(ii) The theory virtually involves that all chemical reactions fall into an acid-base category, and one begins to wonder tit the purpose of using any name other than chemical reaction.

Other theories. Konopic (1949) proposed donor-acceptor acids and bases which amount to little more than that of Usanovich concept. Bjerrum (1951) considered acid as a proton donor while a conventional Lewis acid he referred to as an antibase. Mulliken (1952) developed quantum mechanical treatment of acids and bases. But all these theories add nothing new to our knowledge. Actually we may say that each approach is correct as far as it goes and no conflict exists around them. But one should know the main theme of each concept and should adopt his thinking to the particular problem in hand.

THE SHAB PRINCIPLE

Based on the preferential bonding, Arland, Chatt and Davies (1958) categorised the metal ions into two classes

1. **Class (a) :** This includes ions of alkali metals, alkaline earth metals, lighter transition metals in higher oxidation states like Ti^{4+}, Cr^{4+}, Fe^{3+} and Co^{3+}, and the hydrogen ion. H^+ These ions have the following characteristic features

   (i) They have small size, high polarising power and high oxidation state.

   (ii) Their outer electrons or orbitals are not easily distorted

1. **Class (b) :** This includes ions of the heavier transition metals and those in lower oxidation states such as Cu^+, Ag^+, Hg_2^{2+}, Pd^{2+}, Pt^{2+}, etc. These ions have the following characteristic features

   (i) These are large sized.

(ii) Their outer electrons or orbitals are easily distorted.

Depending upon their behaviour towards the metals of classes (a) and (b), the ligands have been categorised by Arland, Chatt and Davies into the following two classes:

2. **Class (a) :** This includes ligands which preferentially combine with metal ions of class (a) For example, the ligands NH_3, R_3N, H_2O and F^- ions have great tendency to co-ordinate with class (a) metal ions. The tendency of the complexation of the ligands with class (a) metal ions follows the order given below

$$F > Cl > Br > I$$

$$O > S > Se > Te$$

$$N > P > As > Sb$$

2. **Class (b) :** This includes ligands which preferentially combine with metal ions of class (b). For example, ligands such as R_3P (phosphines) and R_2S have great tendency to co-ordinate with class (b) metal ions. The tendency of the complexation of the ligands with class (b) metal ions follows the order given below:

$$F < Cl < Br < I$$

$$O \leq S < Se = Te$$

$$N \leq P < As < Sb$$

Pearson (1963) suggested that terms hard and soft can be used for class (a) and class (b) respectively. Thus, in his classification, metal ions of class (a) are called hard acids and ligands of class (a) are called hard bases. On the other hand, metal ions and ligands of class (b) are called soft acids and soft bases respectively.

Pearson's Principle : The typical Lewis acid-base reaction can be generalized as

$$\underset{\substack{\text{Lewis acid}\\ \text{(Acceptor)}}}{A} + \underset{\substack{\text{Lewis base}\\ \text{(Donor)}}}{:B} \longrightarrow \underset{\text{Complex}}{A:B}$$

To broaden the scope of the reactions we assume that the bond in AB can be ionic, polar, or non-polar. It would be very beneficial if one could predict the stability of the complex A:B.

A concept designed to explain it is called the principle of Soft and Hard acids and bases and was proposed by Pearson (1963). Soft bases have donor atoms which are easily polarised and which have low electronegativity. These criteria are interdepedent, although not absolutely equivalent. They each relate to the ease with which electrons can be distorted away from the donor site.

Hard bases have properties opposite to those of the soft bases : the donor atoms tend to retain its electrons. Thus, hard bases have donor atoms with low polarizabilities and high electronegativities. Within a group in the periodic table, softness increases with increasing size of the donor atom. Thus, of the halide ions., F^- is the hardest and I^- is the softest. Hard base is hard to oxidize and soft base is easy to oxidize.

Soft acids are those which have acceptor atoms large in size, of low positive charge and containing unshared pair (if electrons (p or d electrons) in their valence shell, *i.e.*, they have high polaritability and low electronegativity and do not have a noble gas configuration. Hard acids are those in which acceptor atoms are small in size, of high electronegativity and low polarizability and a noble gas configuration.

Soft acids form stable complexes' with bases that are highly polarisable, and are good reducing agents and not necessarily good bases towards the proton. Hard acids. On the other hand, usually form stable complexes with bases that bind well with protons.

Table 3: Classification of Bases

Hard Bases	*Borderline Bases*	*Soft Bases*
NH_3,RNH_2,N_2H_4	$C_6H_5NH_2$, C_5H_5N,	H^-
	N_3^-, N_2	
H_2O,OH^-,O^{2-}, ROH,	NO_2^-,$SO_{3^{2-}}$	R^-,C_2H_4,C_6H_6,CN^-, RNC, CO
RO^-, R_2O		
CH_3COO^-, CO_3^{2-}, NO_3^-,	Br^-	SCN^-, R_3P, $(RO)_3P$, R_2As
PO_4^{2-}, SO_4^{2-}, NO_2^-		R_2S, RSH,RS^-, $S_2O_3^{2-}$ I^-
F^-, (Cl^-)		

We now apply the idea of soft and hard acids and bases to the question of the stability of the complex A : B The complex AB is most stable when A and B are either both soft or both hard. The complex is least stable when one of the reactants is very hard and the other is very soft. Pearson's principle is only an approximate qualitative prediction of the relative stability for the adducts and is not a theory or any explanation of the observations.

Theories of Hardness and Softness : Several theories have been put forward to explain the fact that the co-ordination compounds of soft acids and soft bases are most stable. The various theories are:

1. **Electronic theory :** According to the theory, hard-hard interactions involve ionic bonding and soft-soft interactions involve covalent bonding. It is expected that small sized and highly charged positive metal ions would he favouring ionic bonding with hard bases of class (a). If hard-soft interaction takes place, the resulting species would be unstable.

Misono and Coworkers (1967) proposed the following equation which correlates hardness and softness.

$$pK = -\log K = aX + bY + c$$

where K is the equilibrium constant for the dissociation of the metal-ligand complex, X,Y are the parameters of the metal ions, a and b are parameters for the ligands, and c is a constant of the ligand to adjust the pK values so that all the values lie on the same scale. For hard acids, the value of Y is found to he > 2.8 and for soft acid > 3.2. For border line acids, the value ranges between 2.8 and 3.2. For example, the values of Y are as follows:

Li^+	Al^{3+}	Mg^{2+}	Na^+	K^+	Ca^{2+}	Fe^{3+}	Co^{3+}	Cs^+	Co^{2+}
0.36	0.70	0.80	0.93	0.93	1.62	2.32	2.56	2.73	2.96

Hard acids

Sn^{2+}	Tl^{3+}	Cu^{2+}	Pb^{2+}	Tl^+	Hg^{2+}	Au^+
3.17	3.2	3.45	3.58	3.78	4.25	5.95

Soft acids

The value of b (ligand parameter) also increases as we move from a hard base to a soft base.

OH^-	NH_3	Cl^-	Br^-	F^-	$S_2O_3^{2-}$
0.40	1.08	2.49	5.5	7.17	12.4

It is interesting to note that a and X are also regarded as another measure of hardness. However, these include the inherent acid base strengths also X has been found to he closely related to the electronegativity of the metal ion and is considered to be measure of the tendency of the metals ions to accept electrons from the ligands.

Further X follows the same order as that William's order for the stability constants of the complexes

$$\text{Mn (II)} < \text{Fe (II)} < \text{Co (II)} < \text{Ni (II)} < \text{Cu (II)}$$

2. **π Bonding theory :** J. Chatt's pi-bonding theory can be applied to explain the bonding between the soft-soft complexes. According to this theory, the soft acids have loosely held outer d-orbitals electrons which can form pi-bonds by donating to suitable ligands. Such ligands are those in which empty d-orbitals are available on the basic atoms, eg. P, As. S, I, etc. The presence of d-orbital, on the ligands (except on CO) helps to strengthen the n-bonding.
3. **Pitzer's theory :** This theory of London and van der Waal's forces explains that the forces between the interacting soft-soft groups depend upon the product of polarizabilities of the interacting groups. Both soft acids and soft base have larges values of polarizabilities.

Electronegativity and Hardness and Softness : Hardness and softness have been found to be related to electronegativity. In brief, highly electronegative species are hard acids while lower electrongative species are soft acids. For example, highly electronegative ions like Li^+, Na^+, etc. are hard acids while transition metal ions having low electronegativities like Cu^+, Ag^+, Cu^+, etc. are soft acids.

Acid and Base Strength of SHAB : Any acid or base may be considered as hard or soft by its apparent preference for the hard and soft reactants. For instance, a base B is considered to be hard or soft depending upon the value of equilibrium constant K for the following reaction:

$$BH^+ + CH_3Hg^+ \rightleftarrows Ch_3HgB^+ + H^+ \quad ...(1)$$

$$BH^+ \rightarrow B + H^+$$

$$k = \frac{[CH_3HgB^+][H^+]}{[CH_3Hg^+][BH^+]} = \frac{[CH_3HgB^+]}{[CH_3][Hg^+][B]} \frac{[H^+][B]}{[BH^+]} \quad ...(2)$$

$$= K_s K_a \quad ...(3)$$

or

$$\log K = \log K_4 + \log K_a \quad ...(4)$$

If the value of K is less than unity, B is termed as a hard base and if it is greater than unity, B is termed as a soft base.

In equation (4), K, denotes the formation constant of CH_3HgB^+ from CH_3Hg^+ and B while K_a denotes the acid dissociation constant of the conjugate acid BH^+. From the values of K and K_a, it is also possible to decide whether B is hard or soft. If pK_a is more than log K_s, B is termed as hard base and if pK_a is less than log K_s, B is termed as soft base.

From the equation (2), it also follows that the reaction (1) will go to the left if B is a hard base. However, it will go to the right, if B is a soft base. Now the question arises : Why methyl mercury (II) cation is selected in the above discussion 7 The simple reasons for this are :

(i) It is a typical soft acid,

(ii) It is univalent like proton, making the calculations simpler.

Hardness and softness arc quite different from the inherent acid-base strengths because these simply refer to the hard-hard and soft-soft interactions. For example, both hydroxide ion and fluoride ion are hard bases. However, the hydroxide ion (whose pK_a is 15.7) is 10^{13} times more basic than the fluoride ion (whose pK_a, is 2.85).

Certain examples are known in which a strong acid or base may displace a weak acid or base although this may violate the principle of SHAB interactions. For example, a stronger soft base sulphite displaces a weaker hard base fluoride from a hard acid proton:

$$SO_3^{2-} + HF \rightarrow SO_3^- + F^-, K_{eq} = 10^4$$

Similarly, a very strong hard base displaces a soft weak base from a soft methyl mercury (II) cation.

$$OH^- + CH_3HgSO_3^- \rightarrow CH_3HgOH + SO_3^-, K_{eq} = 10$$

The above replacement reactions take place against the hardsoft principle. However, when both the strengths and hardness are taken into cansiderations, the hard-soft rule works well. For example, consider the following reactions:

$$CH_3HgF + HSO_3^- \rightarrow CH_3HgSO_3^- + HF\ (K_{sq} = 10^3)$$

$$CH_3HgOH + HSO_3^- \rightarrow CH_3HgSO_3 + H_2O\ (K_{sq} = 10^7)$$

In the above reactions both sulphite ($pK_a = 14.2$) and hydroxyl ion ($pK_a = 15.7$) combine with protons. But the pK_s for sulphite ions with methyl mercury (II) cation is 21.2 while in case of hydroxyl ion is only 9.4 Therefore, sulphite is considered to be a soft base.

Applications' of the SHAB principle : The principle of soft and hard acid-base finds applications in various domains of chemical reactions. Some of them are:

(i) The simplest reaction that can be explained on this concept is

$$LiI + CsF \rightarrow LiF + CsI$$

This is an exothermic reaction with 13.9 kJ/mole heat of reaction. Here soft iodide prefers to combine with the soft caesium ion. The hard fluoride ion combines with hard lithium ion.

(ii) For heterogeneous catalysis, SHAB principle says that soft metals adsorb soft bases. For solubility, the rule is that hard solutes dissolve in hard solvents and soft solutes dissolve in soft solvents.

(iii) *Occurrence of minerals :* The kinds of metal ores found in the earth's crust can be rationalized by using this principle Thus, hard acids such as Ca^{2+}, Mg^{2+} and Al^{3+} appear as $CaCO_3$, $MgCO_3$ and Al_2O_3 respectively. The anions CO_3^{2-} and O^{2-} are also hard.

These three hard acid cations are never found as their sulphides since S^{2-} is a soft base. However, soft acid such as Cu^+, Ag^+ and Hg^{2+} are found to be combined with the soft base S^{2-} as sulphides. The intermediate acids such as Ni^{2+}, Pb^{2+}, and Cu^{2+} are found as both sulphides and carbonates.

(iv) In both BF_3 and BH_3, the boron is trivalent but quite different behaviour can be explained on the basis of SHAB principle. The presence of hard fluoride ions in BF_3 makes it easy to add other hard bases, eg. $H_3N \rightarrow BF^3$ the presence of soft hydride ions in BH_3 makes it easy to add other soft bases.

In BF_3 the boron is nearly B^{3+} . But the soft hydride ions donate negative charges extensively to boron, so that the boron atom is almost neutral and becomes soft.

(v) For elements of variable valency there is usually an increase in hardness with increase in oxidation state. The nickel (0) in. Ni $(CO)_4$ is soft, mickel (II) is borderine and nickel (IV) is hard.

(vi) The sulphur end of the thiocyanate ion is assumed to be softer than the nitrogen end, and hence prefers soft Lewis acids.

(vii) *Relative stabilities of Compounds :* By using Tables I and II and Pearson's rule, it becomes possible to explain the relative stabilities of the compounds and complexes. For example, AgI_2^- is stable, but AgF_2^- does not exist. As described earlier, Ag^+ is a soft acid F^- is a hard base and I^- is a soft base. Therefore, AgI_2^- (soft acid + soft base) is a stable complex and can explain that CoF_6^{2-} (hard acid + hard base) is more stable than CoI_6^{2-} (hard acid + hard base).

By using the SHAB principle, it is possible to predict the relative strength of halogen acids in aqueous medium. For example.

$$HX + H_2O \rightarrow H_3O^+ + X$$

(where X = F , Cl , Br or I)

In the above reaction, the hard acid H+ combines with the hardest base F to form the dissociated or weakest acid. Thus, the strengths of the halogen acids will follow the following order:

$$\underset{\text{Weakest acid}}{HF} < HCI < HBr < \underset{\text{strongest acid}}{HI}$$

Out of the two complexes, $[Cd(NH_3)_4]^{2+}$ and $[Cd(CN)_4]^{2-}$, latter should be more stable (combination of soft acid Cd^{2+} with soft base (CN^-) on the basis of SHAB principle.

(iii) *Poisoning of metal catal catalysts :* This has been explained on the basis of SHAB principle. Soft metals (e.g Pd and Pt) act as catalysts and they are easily poisoned with CO, olefins, phosphorus or arsenic ligands (all soft bases). These ligands adsorb strongly on the surface of the metal and block the active sites. The soft metal catalysts are not affected by hard bases such as F.O.N.

(ix) *Rates of chemical reactions :* The SHAB principle has been used to correlate the rates of chemical reactions. The rates of electrophilic and nucleophilic reactions have been found to depend upon the hardness or softness of the acid or base.

(x) *Course of reactions :* The SHAG principle has been used to predict the course of a reaction. Consider the following reactions:

$$H^+ + CH_3HgOH \rightarrow H_2O + CH_3Hg^+ \quad \text{... (1)}$$

$$H^+ + CH_3HgSH \rightarrow H_2S + CH_3Hg^+ \quad \text{... (2)}$$

According to SHAB principle, the reaction (1) should move towards right because the hard acid H^+ is combining with hard base OH^- the reaction (2) should move towards left because the soft methyl mercury CH_3Hg^+ acid is combining with soft HS^- base.

Limitations of SHAB Concept

The various limitations of this concept are as follows:

(1) It does not involve any quantitative scale of measurement.

(ii) In order to explain the observed phenomenon, one has to break a compound to acid and base fragments. This breaking will be correct only if the reaction is known. For example, the break I for the scarification reactions explains the observed behaviour. However, the break 11 cannot be ignored and should be taken into consideration.

$$CH_3CO^+OH^- \rightarrow C_2H_5O^-H^+ \text{ Break I}$$

$$CH_3COO^-H^+ \rightarrow C_2H_5^+OH^- \text{ Break II}$$

(iii) In this concept, it is assumed that hard-soft factors do not depend upon the acidic or basic character of the compounds and therefore they work independently. However, many examples are known which show interdependence of the two concepts.

(iv) Due to the soft-soft (CH_3^+–H^-) interactions, it is expected that the following reaction should proceed.

$$CH_3^+(g) + H_2(g) \rightarrow CH_4 + H^+(g)$$

However, the unfavourable entropy change (3600 kJ mole^{-1}) indicates that this reaction should not proceed at all and the reason for this is that there is greater acidity of the proton relative to the CH_3 nucleus.

Conclusion

The hard-soft classification is a useful concept, but it does not lead directly to a scale of acid-base strength. Inherent acid-base strengths are also not taken into consideration. Thus the OH^- and F^- ions are both hard bases. But the former is nearly 10^{13} times as strong a base as the latter. Various approaches to correlate inherent acid-base strength with attributed hardness (and softness) factors are yet to be widely accepted.

Interpretation of many reactions by splitting the participants into acid-base fragments is also arbitrary to some extent. The very common reaction between ethanol and acetic acid may be interpreted in either of two ways:

$$CH_3COO^-H+ + C_2H_5 + OH^- \text{ or } CH_3CO + OH^- + C_2H_5O^-H^+$$

Either interpretation is justifiable by the hard-hard combination of H^+ with OH^-. The actual involvement of the second mechanism has to be understood by other means.

Sometimes the hard-soft principle fails to keep parity with inherent acid-base strengths. The reaction:

$$CH_3^+ (g) + H2(g) \rightarrow CH_4(g) + H^+(g)$$

should be favourable in view of the soft-soft combination of CH_3^+ with H^-. But actually the combination in endothermic by about 360 kJ per mole. (This has to be explained by the greater acidity of H^+ relative to CH_{3+}).

Similarly, hard-soft combinations take place in many cases like

$$SO_3^{2-} + HF \rightarrow HSO_3 + F^-$$

Here the soft bane SO_3^{2-} replaces the hard base F^- and combines with the hard acid H^+.The soft SO_3^{2-} is thus stronger than the hard F^-.

However, we should restrict our expectations in view of the statement made by Pearson. "It should be stressed that the SHAB principle is not of theory but is a statement about experimental facts. Accordingly an explanation of some observation in terms of hard and soft behaviour does not invalidate some other, theoretical explanation."

STRENGTH OF HYDRACIDS

The relative strengths of the hydracids of Group VI A and VII A have been found to be as follows:

(i)	H_2Te >	H_2Se >	H_2S >	H_2O
	3	4	7	16
(ii)	HI >	HBr >	HC1 >	HF
	–10	–9	–7	3

The above trend is also found in hydracids of Group V A. For example, the pK_a for ammonia (35) is more than that for phosphine (27).

From the above trend it appears that the strength of the acid is decreasing with the increase in the electronegativity of the elements. However, this trend is not found in the ionic character of the bonds because the ionic character has been found to increase with the electronegativity of the element.

Coordination Chemistry

The branch of inorganic chemistry that deals with the study of coordination compounds is called *coordination chemistry.*

Complex Ion

Complex ion is a positively or negatively charged species which contains a central metalic atom and a suitable number of ligands surrounding the central metallic atom. The central metallic atom may be in zero, positive or negative oxidation state. The ligands may be either neutral molecules or anions (generally) or a combination of the neutral molecules and anions. Examples of complex ions are $[Co^{3+}NH_3)_6]^{3+}$, $[Ni^0(CN)_4]^{-4}$, $[Fe^{2+}CN)_6]^{4-}$, $[Co^-(CO)_4]^-$, $[Co^{3+}(NH_3)_5Cl]^{2+}$ etc.

It may be seen that, depending on the nature of the charge on the complex ion, complex ion may be *cationic complex ion* or *anionic complex ion.* Cationic complex ion has positive charge on it while anionic complex ion has negative charge. Thus $[Co^{3+}(NH_3)_6]^{3+}$ and $[Co^{3+}(NH_3)_5Cl]^{2+}$ are cationic complex ions and $[Ni^0(CN)_4]^{-4}$, $[Fe^{2+}(CN)_6]^{4-}$ and $[Co^-(CO)_4]^-$ are anionic complex ions.

Neutral Complex

A complex which has no charge on it is called neutral complex. It is a non electrolyte, does not undergo ionisation and hence does not give any ions in aqueous solution. Examples of neutral complexes are $[Co^{3+}(NH_3)_3Cl_3]^0$, $[Ni^0(CO)_4]^0$, $(NH_3)_2Cl_2]^0$, $[Ni^{2+}dmg)_2]^0$ etc.

When solutions of two or more simple stable salts are mixed together in simple molecular proportion and the solution thus obtained is allowed to evaporate, crystals of a new compound are obtained. This new compound is called addition or molecular compound. The formation of some addition compounds has been shown below :

Simple compounds		*Addition compounds*
$KCl + MgCl_2 + 6H_2O$	$\rightarrow$	$KCl.MgCl_2.6H_2O$ Carnollite

$K_2SO_4 + Al_2(SO_4)_3 + 24H_2O \rightarrow K_2SO_4.Al_2(SO_4)_3.24H_2O$
Potash alum

$(NH_4)_2SO_4 + FeSO_4 + 6H_2O \rightarrow FeSO_4.(NH_4)2SO_4.6H_2O$
Mohr's salt

$NiCl_2 + 6NH_3 \rightarrow NiCl_2.6NH_3$
Hexammine nickel chloride

$CuSO_4 + 4NH_3 \rightarrow CuSO_4.4NH_3$
Tetrammine copper sulphate

$Fe(CN)_2 + 4KCN \rightarrow Fe(CN)_2.4KCN$
Potassium feirocyanide.

COMPOUNDS FORMATION BY INERT GASES

Earlier attempts to prepare any compound of rare gases were met with failure. The main reasons for this are:

(i) *The complete pairing of all electrons present and the absence of any bonding orbitals.*

(ii) *The existence of these atoms in stable energy states as shown in the Pauling type energy level diagram of these elements.*

(iii) *High ionisation potentials and negligible electron affinity characterising these elements.*

$$He_2 = \overset{1s^2}{\boxed{\uparrow\downarrow}}$$

$1s^2$	$2s^2$	$2p_x^2$	$2p_y^2$	$2p_z^2$
↑↓	↑↓	↑↓	↑↓	↑↓

In spite of the above mentioned unfavourable conditions, certain examples of compound formation have been reported. However, the compounds are formed under unusual conditions.

1. Compound Formation Under Excited Conditions

If we unpair the $1s^2$ electron in the helium atom and promote it to 2s state, then a condition of chemical activity is expected. But this process requires some 1932 kJ of energy per gm atom and can only be realised spectroscopically or under conditions of electric discharge of electron bombardment.

$$He = 1s^2 \text{ (normal, inactive)}$$

$$He = 1s^1 2s^1 \text{ (excited, active)}.$$

In actual practice, some compound formation under excited conditions' takes places. Some well established examples are:

(i) *In a discharge tub,* He_2^+, HeH^+ *and* HeH^{2+} *have been reported but these survive only momentarily.*

(ii) *In a glow discharge, a comparative stable mercury halide* ($HgHe_{10}$) *has been reported to be formed.*

(iii) *When tungsten is bombarded with electrons in an atmosphere of helium, tungsten halide* WHe_2 *is obtained.*

(iv) *Unstable halides of elements such as bismuth, thallium, indium, zinc, sodium, potassium, etc. have been reported to be formed.*

(v) *Metal electrodes in discharge tubes absorb inert gases to form* Pt_3He, *FeHe and FeAr which are probably the interstitial compounds.*

The compound Pt_3He has been thoroughly studied and is found to be different from platinum metal 1 in the following properties:

(i) Pt_3 He *has a ixed decomposition temperature.*

(ii) Pt_3 He gives *an X–ray diffraction pattern which is characteristic of an amorphous substance.*

(iii) Pt_3He *possesses a different density, solubility and electrical resistance from platinum.*

(iv) Pt_3 He *is not decomposed by aqua-regia whereas Pt is attacked by it.*

2. Compound Formation Through Co–ordination

Noble gases may donate an electron pair to a powerful electron acceptor. If such a condition is realised, it means that the largest inert gas atoms must show this property to the greatest extent. In 1935, Booth and Wilson carried out the study of compounds of argon with boron trifluoride (strong electron pair acceptor). Argon forms following type of compounds with boron trifluoride.

$A.xBF_3$ where $x = 1, 2$ or 3 or 6 or 8 or 16

When $x = 1, 2, 3$, the structures are depicted below:

$Ar \rightarrow BF_3$ 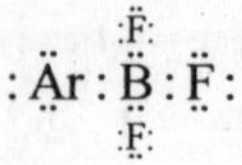 (x = 1)

$BF_3 \leftarrow Ar \rightarrow BF_3$ 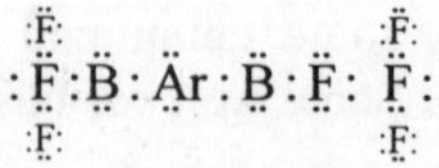 (x = 2)

$BF_3 \leftarrow \underset{\downarrow \atop BF_3}{Ar} \rightarrow BF_3$ 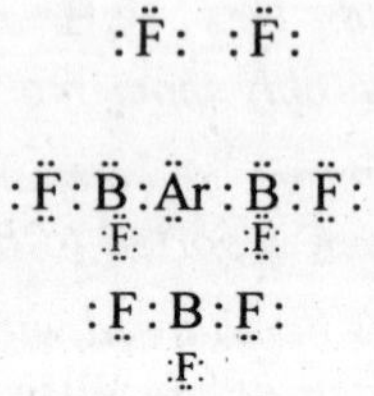

When x = 6, 8 or 16, it was required to postulate that fluorine atom in BF_3 must act as a donor of electron pair to another boron atom giving structure of the type.

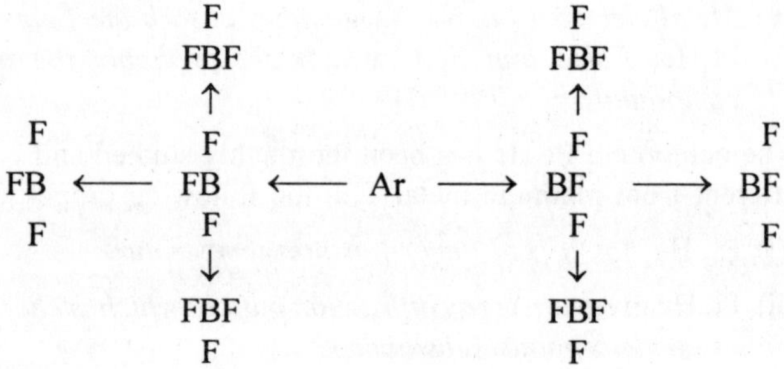

($A.8BF_3$)

In 1918, *Wiberg and Farbe* tried to prepare similar compounds with krypton and xenon which should form more stable compounds because of their greater sizes. They could not isolate such compounds. This led them to the doubt that rare gas atoms could not donate electron pairs to boron. *They repeated the work of Booth* and *Wilson and found that no compound formation occurred between argon and boron trifluoride because argon and boron trifluoride are not even miscible with one another at any temperature.*

3. Compound Formation by Physical Trapping

Until recently, the only well established compounds of the noble gases were those in which gases were trapped in the crystal lattices of other compounds. *These compounds are termed by Powell as* cage compounds or clathrate compounds. *These are so called because one component of the compounds is trapped within the cavities of the crystal lattices of the other component and there is no chemical bond formation between the two components.* These compounds are decomposed on melting or on dissolution of the compounds in any suitable solvent. Examples are:

(i) If quinol (1 : 4 dihydroxy benzene) is crystallised in the presence of heavier noble gases under a pressure of 10 – 40 atmospheres, the gas is trapped in cavities of about 4 Å diameter in the structure. The gas escapes only when the clathrate is dissolved, only Ar, Kr and Xe form clathrate compounds.

Helium and neon fail to form quinol clathrates presumably due to the small sizes of their atoms which enable them to escape from the quinol lattice. Powell confirmed these clathrate compounds of argon, krypton and xenon with quinol of the type:

1 molecule of noble gas: 3 molecules of quinol.

(ii) Similarly, the gases Ar, Kr, and Xe may be trapped in cavities when water solidifies. The so called inert gas hydrates have formulae approximating to $5H_2O$: 1. gas atom.

It is highly probable that clathrates are not chemical compounds.

4. Compound Formation by Dipole-induced Dipole Attraction

The noble gas atoms may be polarised by a strong dipole on other molecules and possess a weak induced dipole of their own. The compounds are expected to be unstable due to weak forces. The solubility of the noble gases in water is high and increases with increased size.

It is probably due to the reason that larger atoms are more easily polarised by the surrounding water molecules. The existence of compounds of argon and krypton with water was confirmed by Tammann and Krige who formulated them as hexahydrates. They also reported $Kr.6D_2O$ and

Tammann and Krig prepared the compounds by compressing the inert gases (Ar, Kr and He) with ordinary water or heavy water at low temperature, and pressure is released. These compounds are called hydrates. The compounds such as $Ar.6H_2O$, $Kr.6H_2O$ and $Xe.6H_2O$ have been obtained and studied.

Hydrates of inert gases are crystalline compounds and their stabilities decrease rapidly from the radon compound down through to argon compound. Similar types of compounds are formed with phenols having the formulae:

$$Kr.2C_6H_5OH$$
$$Xe.2C_5H_5OH.$$

The compounds of inert gases with phenol are isomorphous with the hydrogen sulphide derivative such as $H_2S.\ 2C_6H_5OH$. It is important to remark that lithiurn ions possess tendency to form clusters with inert gases of the type:

$$Li^+nG(n = 1 or\ 2) \text{ where } G = \text{Inert gas}$$

The stabilities of such aggregates and the quantities of gas held increase with the increasing size of the inert gas atoms.

5. Some Recently Discovered Inert Gas Compounds

Only krypton and xenon have been shown to form true chemical compounds. Neither of them were discovered before 1962. Their chemistry still remains limited because the noble gas atoms are only bonded to either fluorine or oxygen in all their stable compounds.

Bartlett, in 1962, found that oxygen gas reacts with PtFe to give a solid compound of the formula O_2PtF_6. The X–ray examination revealed that the compound actually is made up of O_2^+ and PtF_6^- ions and could be represented as $O_2^+PtF_6^-$.

As the ionisation energy of xenon (1176 *kJ*/mol) *is very close to that of oxygen* (1320 *kJ/mol*) *it occurred to Bartlett that the reaction between xenon and* PtFe *should be tried.* He actually found that the reaction took place readily even to room temperature to yield a red solid compound of the composition $XePtF_6$. *This was the first real compound of any of the noble gases.*

The reaction has since been shown to be more complicated, and the product is really $[XeF]^+ [Pt_2F_{11}]^-$.

$$Xe[PtF_6] + PtF_6 \xrightarrow{25°C} XeF]^+[PtF_6]^- + PtF_5$$

$$\xrightarrow{\text{heat } 60°C} [XeF]^+[Pt_2F_{11}]^-.$$

Soon after this it was found that Xe and F_2 reacted at 400°C to give a colourless volatile solid XeF_4. This has the same number of valency electrons as, and is isostructural with, the polyhalide ion $[ICl4]^-$. Following these discoveries there was rapid extension of the chemistry of the noble gases, and in particular of xenon.

The reaction of xenon with PtF_6 demonstrated for the first time that xenon is not a totally inert gas. This led to other investigations in the same direction.

From the ionisation energy values of these elements , it is seen that except radon, which is unstable, only xenon and, to some extent krypton, has ionisation energy which is comparable with common elements such as carbon, nitrogen and oxygen. Xenon thus forms the largest number of compounds. Krypton comes next.

Neon, argon and helium, which have much higher ionisation energies form hardly any compound. Moreover, since reactivity of xenon and krypton is because of their tendency to part with their electrons, it is quite clear that the combining atom should have *strong affinity for electrons, i.e., should be strongly electronegative.* It is due to this that xenon and krypton form compounds mostly with strong electron acceptors, viz., fluorine and oxygen. The most stable of these compounds are *fluorides, oxyfluorides* and *oxides.*

Xenon reacts directly with fluorine. Its compounds with oxygen are obtained.

1. Why noble gas compounds were discovered very late?

The main reason for this is their stable electronic configuration (ns^2 or ns^2p^6) which was regarded as the criterion of chemical inertness. Although existence of compounds having more than eight valency electrons were soon confirmed *e.g.,* PCl_5(10), SF_6(12), IF_7 (14), etc. no serious attempt was made to over–rule the stable octet as the criterion of inertness in noble gas. Some other factors responsible for the late discovery of noble gas compounds are as follows:

(i) Good quality fluorine was not available.

(ii) Fluorine–resistant greases were not available.

(iii) There arises difficulty in handling fluorine gas (non– availability of inert materials to handle fluorine gas).

(iv) Noble gases are not available in plentiful quantities.

2. Why noble gas compounds are common in xenon?

Reactivity of an element is dependent upon its ionization potential; lower the ionization energy of an element, more is its reactivity. Ionization energies of noble gas decrease on moving down the group and thus the reactivity of noble gases gets increased from He to Xe (radon, being radioactive, has not been studied in details).

Table 4

Gas	*He*	*Ne*	*Ar*	*Kr*	*Xe*	*Rn*
I.E., kJ mol^{-1}	2372	2080	1521	1351	1170	1037

Hence **xenon**, which is having the least *I.E.*, forms the largest number of compounds than any other noble gas; krypton forms some compounds but with difficulty as compared to xenon; the lighter members (He, Ne and Ar) do not form any compound.

3. Why do noble gases form compounds only with fluorine and oxygen?

As the chemical reactivity of noble gases involves the loss of electrons, the highly electronegative element (F and O) and group (*e.g.*, $OSeF_5$ and $OTeF_5$ which contain both oxygen and fluorine can) combine with gases very readily. Xe reacts directly only with F_2. However, oxygen compounds can be obtained from the fluorides. There is some evidence for the existence of $XeCl_2$ and $XeCl_4$, and one compound is known with a Xe–N bond. Hence the important noble gas compounds include fluorides, oxyfluoride and oxides of xenon and krypton.

Compounds of Xenon

Xenon and fluorine react under a variety of conditions like temperature, pressure mixing ratio of the constituents, ultraviolet irradiation, electric discharge, etc., to produce xenon fluorides.

For examples:

$$\underset{(2:1\text{ ratio})}{Xe + F_2} \xrightarrow{400°C, 2Hr.} XeFe_2$$

$$\underset{(1:5\text{ ratio})}{Xe + 2Fe_2} \xrightarrow{450°C,\ 475°C} XeFe_4$$

$$\underset{(1:20\text{ ratio})}{Xe + 3F_2} \xrightarrow[50\text{ atm.}]{300°C, 10-20\text{ hr.}} XeF_6.$$

It can be seen from the pressure versus temperature diagrams of xenon fluorides that XeF_4 will always be contaminated with some amounts of lower (*i.e.*, XeF_2) and higher (*i.e.*, XeF_6) fluorides. It can be purified by making use of its inability to donate F^- ion to AsF_5 acceptor. Impure Xep_4 is dissolved in BrF_5 and reacted with an excess of AsF_5.

Table 5 : Structures of some xenon compounds

Formula	*Name*	*Oxidation state*	*m.p. (°C)*	*Structure*
XeF_2	xenon difluoride	(+II)	129	linear (RnF_2 and $XeCl_2$ are similar)
XeF_4	xenon tetrafluoride	(+IV)	117	square planar ($XeCl_4$ is similar)
XcF_6	xenon	(+VI) hexafluoride	49.6	distorted octahedron
XeO_3	xenon	(+VI) trioxide	explodes	pyramidal (tetrahedral with one corner unoccupied)
XeO_2F_2	—	(+VI)	30.8	trigonal bipyramid (with one position unoccupied)
$XeOF_4$	—	(+VI)	–46	square pyramidal (octahedral with one position unoccupied)
XeO_4	xenon tetraoxide	(+VIII)	–35.9	tetrahedral
XeO_2F_2		(+VIII)	–54.1	trigonal bipyramid
$Ba_2[XeO_6]^{4-}$	barium perxenate	(+VIII)	dec. > 300	octahedral

Xenon difluoride and xenon hexafluoride form involatile salts while unreacted BrF_5 is distilled off at 273°C. The unchanged pure XeF_4 is then obtained by vacuum distillation at room temperature. The various characteristics of xenon compounds are listed in Table 5.

Why xenon and bromine, argon and fluorine do not form compounds like xenon fluorides?

Electronegativity of bromine is much less than that of fluorine, so bromine is unable to bring about promotion of 5p electrons of xenon to 5d–orbitals. As the electrons remain paired, no compounds are formed. Ionization energy of argon is quite high, *i.e.,* higher than that of xenon. Therefore, the energy released by Ar–F bonds is insufficient to promote electron from 4p to 4d electrons to give rise to unpaired electrons. As the electrons remain paired, no compounds are formed.

XENONDIFLUORIDE (XEF_2)

Preparation

(i) It has been prepared by irradiating a mixture of xenon and fluorine with light from a high pressure mercury arc.

$$Xe + F_2 \xrightarrow[\text{Mercury lamp}]{\text{Light}} XeF_2$$

(ii) It may be conveniently prepared by circulating a mixture of Xe and' fluorine through a nickel tube at 400°C for about 8 hours.

$$Xe + F_2 \xrightarrow{400°C} XeF_2$$

When the mixture of XeF_2 unreacted Xe and fluorine is fed into a U–tube maintained at – 50°C. XeF_2 is condensed into it.

XeF_2 has to be removed as soon as it is formed otherwise it reacts with more fluorine to form XeF_4. The XeF_2 can be condensed and thus removed from the unreacted xenon and fluorine by passing the reaction product through a U tube maintained at – 50°C.

$$XeF_2 + F_2 \longrightarrow XeF_4$$

(iii) It can be prepared by the reaction of xenon with O_2F_2 at –118°C.

$$Xe + O_2F_2 \xrightarrow{-118°C} XeF_2 + O_2$$

Properties

(i) It forms transparent crystalline solid melting at –127°C.

(ii) It is less volatile than hexafluoride and can be stored in nickel containers for a long time when it is free from impurities.

(iii) It dissolves in anhydrous liquid HF and the solution does not conduct electricity,

(iv) XeF_2 is a mild fluorinating agent. At room temperature, ethylene is fluorinated to form 1,1 and 1,2 – difluoroethanes.

$$XeF_2 + 2C_2H_4 \longrightarrow \begin{matrix} CH_2F \\ | \\ CH_2F \end{matrix} + \begin{matrix} CH_2F \\ | \\ CH_3 \end{matrix} + 2He$$

With propylene, the main product is, 1,1–difluoropropane

$$XeF_2 + C_3H_6 \longrightarrow CF_2HCH_2CH_3 + Xe$$

(v) XeF_2 reacts with oxides of nitrogen. With NO_2 the reaction is slow but with nitric oxide it is rapid and is represented as:

$$XeF_2 + 2NO \longrightarrow 2NOF + Xe$$

(vi) XeF_2 reacts quantitatively with hydrogen as follows:

$$XeF_2 + H_2 \longrightarrow 2HF + Xe$$

(vii) Its hydrolysis proceeds very rapidly in an alkaline medium but slowly in neutral or acid medium.

When it is boiled with 1 molar solution of sodium hydroxide, it undergoes hydrolysis liberating oxygen quantitatively according to the following equation:

$$XeF_2 + H_2O \longrightarrow Xe + \frac{1}{2}O_2 + 2HF$$

At first the hydrolysis produces a bright yellow colour which rapidly disappears and the solution becomes colourless at room temperature.

(viii) XeF_2 forms addition compounds with SbF_5, TaF_5, NbF_5) RuF_5 etc.

$$XeF_2 + 2SbF_5 \longrightarrow XeF_2.2SbF_5$$

(ix) On heating it undergoes disproportionation.

$$2XeF_2 \longrightarrow Xe + XeF_4$$

XeF_2 oxidises Cl^- to Cl_2, I^- to I_2 and cerium (III) to cerium (IV).

$$XeF_2 + 2HCl \longrightarrow 2HF + Xe + Cl_2$$

$$SO_4^{2-} + XeF_2 + Ce_2^{III}(SO)_3 \longrightarrow 2Ce^{IV}(SO4)_3 + Xe + F_2$$

Xenon Tetrafluoride

Preparation

(i) It is prepared by heating a mixture of fluorine and xenon in the ratio of 5 : 1 in a nickel vessel at 450 – 475°C and 6 atmospheric pressure for a few hours. After heating, the reaction is quenched by cooling the nickel vessel rapidly in cold water. The excess of fluorine is removed by pumping at – 78°C.

$$Xe + 2F_2 \xrightarrow{450-475°C} XeF_4$$

The tetrafluoride prepared under the above mentioned conditions contains less than 8% XeF_2 and XeF_6 as impurities. These impurities can be removed from XeF_4 by sublimation.

(ii) It is also prepared by passing an electric discharge through a mixture of xenon and fluorine.

Properties

(i) At room temperature, it forms transparent crystals having m.p.– 114°C.

(ii) It is quite unstable and can be stored in a nickel vessel.

(iii) It is less volatile than hexafluoride.

(iv) It dissolves in trifluoroactic acid.

(v) It is hydrolysed by H_2O fumming XeO_3

$$3XeF_4 + 6H_2O \longrightarrow XeO_3 + 2Xe + \frac{3}{2}O_2 + 12HF$$

This reaction is indeed a major hazard in Xe/F chemistry, since XeO_3 is highly explosive, and the complete exclusion of moisture is therefore essential. Interestingly the maximum yield of XeO_3 is 33% rather than the 50% that would be expected from a simple disproportionation of and the following reaction sequence has been suggested to explain this:

$$3Xe^{IV}F_4 + 6H_2O \longrightarrow \{Xe^{II}O\} + \{Xe^{VIII}O_4\} + 12HF$$

$$\{Xe^{II}O\} \xrightarrow{\text{Decomposition}} 2Xe^0 + O_2$$

$$\{Xe^{VIII}O_4\} \xrightarrow{\text{Decomposition}} X^{VI}O_3 + \frac{1}{2}O_2$$

The stoichiometry of the reaction also depends sensitively on the precise conditions of hydrolysis,

(vi) It reacts with NO, NO_2, HCl and NH_3.

$$XeF_4 + 4NO \longrightarrow 4ONF + Xe$$

$$XeF_4 + 4NO_2 \longrightarrow 4O_2NF + Xe$$

$$XeF_4 + 4HCl(g) \longrightarrow 2HF + 2Cl_2 + Xe$$

$$XeF_4 + 6NH_3(g) \longrightarrow 4NH4F + N_2 + Xe + H_2$$

(vii) XeF_4 is a strong fluorinating agent. It fluorinates ethylene and propylene.

$$2C_2H_4 + XeF_4 \longrightarrow \begin{matrix} CH_2F \\ | \\ CH_2F \end{matrix} + \begin{matrix} CH_2F \\ | \\ CH_3F \end{matrix} + Xe$$

It fluorinates compounds:

$$XeF_4 + 2SF_4 \longrightarrow Xe + 2SF_6$$

$$XeF_4 + Pt \longrightarrow Xe + PtF_4$$

$$XeF_4 + C_6H_6 \longrightarrow Xe + C_6H_5F + HF$$

(viii) When XeF_4 is warmed with excess of xenon, XeF_2 is formed

$$XeF_4 + Xe \longrightarrow 2XeF_2$$

(ix) With platinum, XeF_4 in anhydrous HF gives PtF_4

$$2Pt + XeF_4 \longrightarrow PtF_4 + 2Xe$$

(x) There is some evidence that XeF_4 forms an addition compound of composition $XeF_4.2SbF_5$.

Xenon Hexafluoride (XeF_6)

Preparation

(i) It is very difficult to prepare. However, it is prepared by heating a mixture of xenon and fluorine in the ratio of 1: 20 in a nickel vessel and at 300°C and 50 atmospheric pressure.

$$Xe + 3Fe_2 \xrightarrow[50\ atm.]{300°C} XeF_6$$

The yield of this method is 95%. It becomes 100% when the pressure is. 2 atmospheres and the temperature 700°C.

(ii) It is prepared by the electric discharge in a mixture of fluorine and xenon in 3 : 1 ratio at low temperatures.

(iii) XeF_6 is also formed by the reaction of XeF_4 with O_2F_2 at temperatures between –133 and – 78°C

$$XeF_4 + O_2F_2 \longrightarrow O_2 + XeF_6$$

XeF_6 cannot be handled in glass or quartz apparatus because of stepwise reaction which finally produces the dangerous XeO_3:

$$2XeF_6 + SiO_2 \longrightarrow 2XeOF_4 + SiF_4$$

$$2XeOF_4 + SiO_2 \longrightarrow 2XeO_2F_2 + SiF_4$$

$$2XeO_2F_2 + SiO_2 \longrightarrow 2XeO_3 + SiF_4$$

Properties

It is a colourless crystalline solid at room temperature and melts at 47°C to give a yellow liquid but does not possess definite melting point.

(ii) It is best stored in nickel containers because it rapidly attacks glass and silica,

(iii) It is soluble in anhydrous hydrogen fluoride and is appreciably ionized as

$$HF + XeF_6 \longrightarrow XeF_5^+ + HF_2^-$$

It also reacts with alkali metal fiuorides except LiF to form heptafluorides.

$$XeF_6 + RbF \longrightarrow RbXeF_7$$

The rubidium and caesium salts are unstable and undergo decomposition.

$$2RbXeF_7 \longrightarrow XeF_6 + Rb_2XeF_8$$

(iv) Like other fiuorides, it also gets reduced by hydrogen to xenon and hydrogen fluoride.

$$XeF_6 + 3H_2 \longrightarrow Xe + 6HF$$

(v) It rapidly reacts with HCl and ammonia in accordance with the following equations:

$$XeF_6 + 6HCl \longrightarrow Xe + 3Cl_2 + 6HF$$

$$XeF_6 + 8NH_3 \longrightarrow Xe + 6NH_4F + N_2$$

(vi) When it is treated with water, it undergoes hydrolysis to form XeO_3 or $XeOF_4$.

$$XeF_6 + 3H_2O \longrightarrow XeO_3 + 6HF$$

$$XeF_6 + H_2O \longrightarrow XeOF_4 + 2HF$$

Probably this reaction proceed *via* the intermediate product $XeOF_4$. But this is still not certain.

(vii) On strong heating, it decomposes to give XeF_2 and XeF_4.

(viii) Out of three fluorides of xenon, XeFe is the best fluoronating agent.

(ix) Many addition compounds with fluorides of antimony, tin, arsenic, germanium, etc., are known.

$$AsF_5 + XeF_6 \longrightarrow XeF_6 .\ AsF_5 \text{ or } [XeF_5]+ [AsF_6]^-$$

$$SnF_4 + 4XeF_6 \longrightarrow 4XeF_6.SnF_4$$

$$XeF_6 + BF_3 \longrightarrow XeF_6.BF_3 \text{ or } [XeF_5]^+ [BF_4]^-$$

(x) It reacts with fluoride ion donors to from $[XeFs]^{2-}$ species. For instance.

$$CsF + XeF_6 \longrightarrow CsF.XeF_6 \xrightarrow{CsF} Cs_2XeF_8 \text{ or } 2Cs^+[XeF_8]^{2-}$$

(xi) It reacts with sodium hydroxide to form white crystalline sodium perxenate.

$$XeF_6 + 1\ ONaOH \longrightarrow Na_4XeO_6 + 6NaF + 4H_2O + H_2$$

HYDRATES, AMMONIATES AND PEROXOHYDRATES

Many salts crystallise with solvent molecules like H_2O, NH_3 and H_2O_2 The salts obtained in this way are also *addition compounds*. When the added solvent molecule is H_2O, NH_3 or H_2O_2, the addition compounds formed are called *hydrates*, *ammoniates* and *peroxohydrates*. In naming these compounds the number of solvent molecules is indicated by Arabic numbers (1,2,3 etc.). For example $CuSO_4.5H_2O$ [Copper (II) sulphate 5-hydrate], $CaCl_2.6NH_3$ (Calcium chloride 6-ammoniate), Na02H.H202 (Sodium hydrogen peroxide 1-peroxohydrate).

Types of Addition Compounds : Double Salts and Coordination (or Complex) Compounds

Addition compounds are of two types :

1. *Double salts* : These are the addition compounds which are stable in the solid state but give their constituent ions when

dissolved in water or in any other ionic solvent. In these compounds the individual properties of the constituent ions are not lost.

For example, when carnollite, $KCl.MgCl_2.6H_2O$ which is a double salt gives its constuent ions namely K^+, Mg^{2+} and Cl^- ions when dissolved in water and hence the aqueous solution of $KCl.MgCl_2.6H_2.6H_2O$ gives the tests of K^+ Mg^{2+} and Cl^- ions. In other words in double salts, these are the addition compounds which exist only in crystal lattices but break down into their constituent compounds when these are dissolved in water or any other solvent.

It means that the physical and chemical properties of double salts are essentially the same as those of the individual compounds. For example, potash alum is a double salt. This salt is obtained when potassium sulphate and aluminium sulphate are mixed in stoichiometric amounts in solution and the solution is crystallised.

$$K_2SO_4 + Al_2(SO_4)_3 + 24H_2O \rightarrow K_2SO_4.Al_2(SO_4)_3.\ 24H_2O$$

Potash alum is stable only in the crystalline state. But when dissolved in water, it loses its identity and breaks up into its constituent simple ions, K^+ Al^{3+} and SO_2^{-4}.

Thus, potash alum is an example of double salts. Similarly, carnallite, $KCl.MgCl_2.\ 6H_2O$ is another example. Its aqueous solution shows the properties of K^+, Mg^{2+} and Cl^- ions. Mohr's salt, $FeSO_4.(NH_4)_2SO_4.6H_2O$ is also a double salt because it shows the properties of Fe^{2+}, NH_4^+ and SO_4^{2-}.

$$K_2SO_4 + Al_2(SO_4)_3 + 24H_2O(s) \rightarrow 2K^+ (aq) + 2A1^{3+} (aq) + 4SO_4^{2-}(aq)$$

$$FeSO_4.(NH_4)2SO_4.6H_2O(s) \rightarrow Fe^{2+}(aq) + 2NH_4^+(aq) + 2SO_4^{2-}(aq)$$

$$KCl.MgCl_2.6H_2O(s) \rightarrow K^+ (aq) + Mg^{2+}(aq) + 3Cl^-(aq).$$

2. *Coordination (or complex) compounds :* These are the addition compounds which donot give all their constituent ions when dissolved in water. In these compounds the individual properties of some constituent ions are lost.

   For example $CuSO_4.4\ NH_3$ which is a complex compound does not give all its constituent ions viz. Cu^{2+} and SO_4^{2-} in solution. It gives only SO_4^{2-} ion and Cu^{2+} ion is obtained as $[Cu(NH_3)_4]^{2+}$ ion which is called complex ion.

$$CuSO_4.4NH_3 \rightleftharpoons [Cu(NH_3)_4]^{2+} + SO_4^{2-}$$

Thus $CuSO_4.4NH_3$ is formulated as $(Cu(NH_3)_4]SO_4$. It follows from this discussion that $CuSO_4.4NH_3$ gives the tests of SO_4^{2-} only and does not show the tests of Cu^{2+} ions. $[Cu(NH_3)_4]SO_4$ is called *complex* (or *coordination*) compound. A complex compound may be composed of simple cation and complex anion, complex cation and simple anion or complex cation and complex anion as shown below :

$K_3[Fe(CN)_6]$	$\rightleftharpoons$	$3K^+$ Simple cation	+	$[Fe(CN)_6]^{3-}$ Complex anion
$[Co(NH_3)_6]Cl_3$	$\rightleftharpoons$	$[Co(NH_3)_6]^{3+}$ Complex cation	+	$3Cl^-$ Simple anion
$[Pt(NH_3)_4]\ [PtCl_4]$	$\rightleftharpoons$	$[Pt(NH_3)_4]^{2+}$ Complex cation	+	$[PtCl_4]^{2-}$ Complex anion

A neutral complex compound like $[Ni^0CO)_4]^0$, $[Ni^{2+}dmg)_2]^0$, $[Pt^{2+}(NH_3)_2Cl_2]^0$ etc. does not undergo ionisation and hence does not give any ions in aqueous solution. In other words the coordination compounds are molecular or addition compounds which retain their identity even in solution and the properties of which are different from those of their constituents are called complex salts or coordination compounds.

For example, when a solution of ferrous cyanide is mixed with potassium cyanide, potassium ferrocyanide is formed which in aqueous solution does not give tests for Fe^{2+} and CN^- ions but gives the tests of ferrocyanide ion; $[Fe(CN)_6]^{4-}$.

$$Fe(CN)_2 + 4KCN \rightarrow K_4Fe(CN)_6.$$

$$K_4Fe(CN)_6 \rightleftharpoons 4K{+}{+}Fe(CN)_6{}^{4-}$$

Another example is that when ammonia is added to copper sulphate solution, a deep blue colour due to the formation of tetraammine cupric sulphate is obtained.

$$CuSO_4 + 4NH_4OH \rightarrow [Cu(NH_3)_4]SO_4 + 4H_2O$$

Tetraammine cupric sulphate when dissolved in water does not give the test of Cu^{2+} ions but instead gives $[Cu(NH_3)_4]^{2+}$ ions.

$$[Cu(NH_3)_4]SO_4 \rightleftharpoons [Cu(NH_3)_4]^{2+} + SO_4{}^{2-}$$

From the above examples, it is evident that $K_4Fe(CN)_6$ and $[Cu(NH_3)_4]SO_4$ are co-ordination compounds.

Coordination compounds are known by various names such as *complex compounds*, *Werner complexes* and *additive compounds*. The complex compounds formed by the association of neutral molecules of ammonia with metal cations are called *metal ammines*.

Coordination compounds occur in plants as well as in animals. Chlorophyll which is an important constituent of plant life is a coordination compound containing Mg(II). The colour of blood is due to haemoglobin which is a coordination compound of Fe(II). Most of other trace elements in human body function in the form of coordination compounds.

Nomenclature of Double Salts

(a) Double salts are named by connecting the names of the individual compounds by hyphens (–). The number of the molecules of individual compounds is indicated by using *Arabic numerals* (*e.g.*, 1, 2, 3 etc.) separated by solid lines and enclosed in brackets.

Some Examples:

$K_2SO_4.Al_2(SO_4)_3.24H_2O$ ... Potassium sulphate-aluminium sulphate-water (1/1/24)

$FeSO_4.(NH_4)_2SO_4.6H_2O$ ... Ferrous sulphate ammonium sulphate-water (1/1/6)

$Na_2CO_3.10H_2O$ ... Sodium carbonate water (1/10)

$3CdSO_4.8H_2O$ ... Cadmium sulphate water (3/8)

$AlCl_3.4C_2H_5OH$ ... Aluminium chloride-ethanol (1/4)

$NH_3.BF_3$ ... Ammonia-boron trifluoride (1/1)

$8H_2S.4_6H_2O$ ... Hydrogen sulphide water (8/46)

(b) If a double salt contains the cations in the same oxidation state, the cations are written alphabetically. Hydrogen is an exception which is written in the end. For example, $KHCO_3$ (Potassium hydrogen carbonate), $KNaCO_3$ (Potassium sodium carbonate), $NaNH_4HPO_4.4H_2O$ (Ammonium sodium hydrogen phosphate 4-hydrate) etc.

(c) If a double salt contains the cations in different oxidation states, the cations are written in the increasing order of their oxidation states, hydrogen being an exception which is written at the end.

Examples are $KMgCl_3.6H_2O$ (potassium magnesium chloride 6-hydrate), $KA1(SO_4)_2.12H_2O$ (potassium aluminium sulphate 12-hydrate), $NH_4MgPO_4.6H_2O$ (ammonium magnesium phosphate 6-hydrate), $(NH_4)_2CO(SO_4)_2.6H_2O$ [ammonium cobalt (II) sulphate 6-hydrate], $NaZn(UO_2)_3(CH_3COO)_9.6H_23$ (sodium zinc triuranyl acetate 6-hydrate).

(d) If a double salt contains the anions in the same oxidation state, the anions are written alphabetically. For example PbClF is called lead (II) chloride fluoride.

(e) If a double salt contains the anions in different oxidation states, the anions are arranged in the order : O^{2-}, OH^-, other simple inorganic anions with two elements, organic anions, H^-.

The number of ions is indicated by writing the prefix bis. iris etc. in parentheses. For example $NaCl.NaF.2Na_2SO_4$ is called (hexa) sodium chloride fluoride (bis) sulphate.

(f) The hydrolysed or basid: salts are also treated as double salts containing more than one anion and hence are named by the same rules as used for naming double salts. For example BiOCI [bismuth (III) oxide chloride], Sn(OH)CI [tin (IV) hydroxide chloride], $ZrOCl_2.SH_2O$ [zirconium (IV) oxide (di)chloride 8-hydrate], $CuCl_2.3Cu(OH)_2$ or $Cu_2(OH)_2Cl$ [(di) copper (II) trihydroxide chloride].

Note: It may be noted that for H_2O molecule we have used *hydrate* and the number of H_2O molecules has been indicated by Arabic numbers.

CLATHRATE COMPOUNDS

The first known *clathrate,* discovered in 1811, was a gas hydrate of chlorine, which *Sir Humphrey Davy* believed to be a true chemical compound. Organic clathrate compounds have been known since 1849 and by 1900, a number of these species had been prepared.

The nature of these materials was not understood until the X–ray work of Powell and his co–workers provided detailed structural information on a variety of examples. However, Powell defined the clathrate compounds as given below:

"Noble gases form compounds in which these gases are trapped within the cavities of crystal lattices of some inorganic and organic compounds. The compounds so formed are called clathrate compounds."

Powell showed that these compounds consisted of atoms or molecules of one substance called the *guest'* held in crystalline cages of another substance called the *'host'*.

The 'guest' molecules are prevented from leaving the host cages by repulsive forces acting between the electron clouds of molecules forming the cavities, *i.e.,* '*host*' and of *'guest'* molecules. *The* '*guest*' can only escape from the 'host' when the forces holding the molecular cages together are overcome.

Why Name Clathrate?

Powell named these substance '*clathrates*' from the Latin word '*clathrates'* which means enclosed or protected by crossbars grating.

Modem Definition

The definition given by Powell is not general one and limited to inert gases only. However, it is observed that many clathrate compounds are known in which inert gases are not present. In its narrowest definition.

"A clathrate is a single phase solid in which atoms or molecules of a guest component are completely surrounded and kept as in a cage by the crystalline lattice of host component without any sensible interaction between them except van der Wools forces."

Many authors have broadened the above definition to include compounds in which the guest component is held as in an open channel or even as in a sandwich by the crystalline lattice of the host component. The crystalline lattice of the host component is generally much more expanded in its clathrate from that in pure form.

Stability of Clathrate Compounds

The main reasons are:

(i) The stability of many clathrate compounds is due to the formation, in the expanded lattice of the host itself, of strong interactions such as hydrogen bonds which do not exist to such an extent in the normal crystalline lattice of the host alone.

(ii) van der Waals forces and even sometimes weak charge transfer interactions contribute to the stability of clathrates.

(iii) Steric factors are always of prime importance because the clathrated molecule must adopt to the size and the shape of the rigid cage formed by the crystalline lattice of the host.

CHARACTERISTICS OF CLATHRATE COMPOUNDS

(i) These compounds are decomposed on melting or on dissolution of the compound in any suitable solvent.

(ii) Clathrate compounds cannot be considered to be compounds, since there are no specific chemical bonds between the *'guest'* and the '*host*' molecules, but there are only weak intermolecular forces.

(iii) According to X–ray studies, the lattice structure of the host in the clathrate is not in its normal crystalline form.

(iv) The clathrate form of the host lattice when its cavities are not occupied by guest atoms or molecules is usually less stable than the normal form.

(v) A clathrate compound is only formed when the guest molecules have such size and shape to fit the geometries of the cavities of host.

(vi) Although, the clathrates are not stoichiometric compounds, yet they can have a maximum composition formula.

$$nH.mG$$

In which m is the maximum number of guest atoms (G) or molecules which can be accommodated in a π cage of host molecule (H).

EXAMPLES OF CLATHRATE COMPOUNDS

A few clathrates that have historical, theoretical or practical importance will be used to illustrate the subject.

Quinol–Noble Gases Clathrates

Quinol (co–dihydroxybenzene) when crystallises from its solution in water or alcohol under a pressure of 20 atmospheres of a noble gas (argon, krypton or xenon) forms crystals of quinol-argon, quinol– krypton or quinol–xenon clathrate compounds. These quinol gas clathrate compound crystals differ form ordinary quinol crystals in their form and size. The noble gas in these quinol noble gas clathrates compounds is enclosed or entrapped in the crystal lattice of the quinol. *The properties of quinol- noble gas clathrates are as follows:*

(i) The crystals of these clathrates are quite stable. They can remain as such for several years.

(ii) On heating or dissolving in water, the noble gas is expelled leaving behind quinol as such. X–ray studies reveal that in crystal lattice of quinols, three molecules form a special cage of diameter 4 Å (4×10^{-8}cm).

In the quinol crystal, there are many such cavities connected together through very fine apertures within the crystal. Quinol, when crystallised from its aqueous or alcoholic solution in presence of other compounds, the atoms or molecules of which can fit tightly in its cavities form clathrate compounds. The entrapped molecules cannot come out.

(iii) The ratio of cavities to quinol molecules in the clathrate structure is 1:3 and the maximum composition formula expected is $[C_6H_4(OH)_2]_3Ar$. A total filling of the cages (I argon for 3 quinol molecules) is never reached during the crystallisation of quinol.

Structure of Quinol–Noble Gas Clathrates

Powell in England (1948) resolved the structure of these clathrates by X–ray diffraction method. In these clathrates all the hydroxyl groups are tied together through hydrogen bridges, forming hexagons with oxygen atoms at each angle and the benzene rings pointing alternatively up and down, the interlocking of all hydroquinone molecules form an infinite three-dimensional cage work, the voids of which have a diameter of 7.6 A°, big enough to entrap argon atoms.

In 1957 van der Waals predicted a neon/hydroquinone clathrate which might be possibly formed at pressure in excess of 160 atm. The first evidence for such a species was obtained by Gnauk (1964). The clathrate was obtained from a sorbitol/quinol solution and contained 4 wt percent of neon. Gnauk showed that all quinol clathrates crystallised from sorbitol contain higher noble gas concentrations than those crystallised from aqueous solutions. The structures also differ.

Helium does not form quinol clathrate. The atoms or molecules which are to be entrapped must be of suitable size (diameter of cavities in quinol's crystal lattice is approximately 7.9Å). The molecules should neither be too large (as they will not fit in) nor too small (as they will easily escape out). Helium atoms are believed to be too small and too difficult to deform or polarise (as required for van der Waals bonding) to be imprisoned in quinol cage.

Hydrates

(a) General Information

Villard and de Forcard (1925) prepared the first hydrates of argon, krypton, and xenon. Broadly speaking the cavities where the guest molecules are held in hydrates can be regarded as spherically symmetrical. There are tow structures I and II which are found in gas hydrates. In both structures the pentagonal dodecahedron is an essential unit. The dodecahedron exhibits degrees of distortion from their ideally regular forms.

In both structure the pentagonal dodecahedron is an essential unit. The dodecahedra exhibit degrees of distortion from their ideally regular form. Structure I contains 46 water molecules in the unit cell; there are two small cavities and six larger cavities. The small cavities are bounded by twenty water molecules while the larger cavities are enclosed by 24 water molecules. The small and large cavities have free diameters of about 5.1 and 5.8 A° respectively and each cavity is capable of holding one guest. This gives rise to the limiting theoretical composition as $5.75H_2O$. IG.

Structure II hydrates contain one hundred thirty six water molecules in the unit cell and there are sixteen small dodecahedral and eight larger hexadecahedral cavities. Their free diameters are 5.0 and 6.7 Å respectively. The large cavities often permit the formation of this structure type when the guest molecules are too large for the cavities in structure I when the hexadecahedral cavities are singly occupied, the limiting composition is $1.7H_2O$.IG.

Types of Hydrates

(*i*) *Simple Hydrates.* Some simple noble gas hydrates are formed if water is crystallised slowly in the presence of the gases. No helium or neon hydrates have been prepared, even under high pressure, presumably because the helium and neon atoms are too small and too volatile to be retained in the cavities. Argon, krypton and xenon, however, form single-phase crystals of a structure I type in which $n = 46$ and $m = 8$ in the maximum composition formula $nH_2O.mNg$ where Ng is the noble gas.

Experimentally a ratio of $6H_2O$; 1 Ng is found which is very close to the theoretical value. When water is allowed to freeze in the presence of noble gas (argon, krypton and xenon) under pressure, crystalline, hydrates are formed. The number of water molecules per noble gas is nearly 6.

Ar. $5H_2O$; Kr.$6H_2O$; Xe.$6H_2O$

These noble gas hydrates are stable at very low temperatures. At higher temperatures, the noble gas escapes out. In these hydrates the covalently bonded water molecules are held together by hydrogen bonds. There is relatively little interaction between the noble-gas guest atoms and the water molecules cages.

(c) Helium and Neon do not Form Hydrates

(i) Helium and neon do not form hydrates, even under high pressure. Non–formation of helium and neon hydrates may probably be because of the fact that *helium and neon atoms are too small and too volatile to be retained in cavities.*

(ii) *Hydrates with heavy water.* Hydrates of krypton and xenon with heavy water have also been prepared.

(iii) *Double Hydrates.* Double hydrates are formed in the presence of small atoms (such noble gases) or molecules (such as H_2O, O_2, N_2, CO_2) then double hydrates are formed. These contain water, a large molecule (organic compounds like acetone, chloroform, carbon tetrachloride etc.) and a small molecule (noble gases). The general formula of each is:

$$\text{org. 2G. } 17H_2O$$

Org. = Molecule of acetone, chloroform, carbon tetrachloride or methylene chloride.

G = Noble gas (Ar, Kr, Xe).

In fact, the most remarkable aspect of these materials is their high decomposition at 1 atm. pressure. In each case the xenon containing double hydrate is the most stable because on account of the volatility of xenon occupies a greater proportion of the cavities and thereby has a greater stablising influence.

Radioactive Clathrates

Chleck and Ziegler (1959) prepared the krypton/hydroquionone clathrate from krypton containing 5 wt per cent of radioactive ^{85}Kr. Their final product had an activity of about three curies per gm. This was the first ^{85}Kr clathrate prepared. It is now known that krypton–85 can be stably incorporated into various solids and these are called *kryptonates. The* release of ^{85}Kr from the solid proceeds in proportion to the rate of any reaction occurring at the surface of the solid. This effect has been used to study reactions at the surface of a solid, such as corrosion rates, and for the analytical estimation of reactive gases.

The greatest application of kryptonates appears to have occurred in the realms of analytical chemistry. A quinol/Kr^{85} clathrate was prepared by heating solid quional to 186°C under a pressure of ^{85}Kr and allowing the mixture to cool slowly. The decomposition of this clathrate by ozone and chlorine dioxide was studied. Analysers suitable for the determination of ozone and atomic oxygen in the atmosphere have been devised by Chleck (1962). These system employ the hydroquinone/Kr^{85} clathrates.

Phenol Clathrates

Williams (1959) prepared the phenol clathrates of argon, krypton and xenon by direct combination of the gas with crystalline phenol and in the case of krypton and xenon with molten phenol. These results indicated that the ratio of argon and krypton to phenol may be 1 : 4 rather than 1 : 3 as predicted by Von Stackelberg (1958). The work done by Williams has produced a number of anomalies which still remain unresolved.

Complex Clathrnates

Three principal types of nickel Werner complexes form clathrates, but only with aromatic molecules. This fact suggests that van der Waals forces are not the only one to contribute to the stability of such clathrates but in some cases interactions of the charge transfer type between the host and guest molecules have been clearly demostrated. In such compounds, the clathrate character remains the most important because for molecules having the same charge transfer potential, it is their shape which is decisive for clathrate formation.

Now we will discuss the three types of such complexes;

(i) *Cyano momoammine nickel II clathrates.* The general formula of such clathrates of $Ni(CN)_2NH_3.M$ where M can be benzene, thiophene, furan, pyrrole, aniline or phenol.

In its clathrate form this complex crystallises in planar arrays of nickel cyanide: the ammonia molecules, bonded to nickel atoms, project above and below each plane, delineating cavities between them; molecules larger than that mentioned above cannot be accommodated in the cages offering a purification process for the latter.

An ammoniacal solution of nickel cyanide productes a pale coloured clathrate when shaken vigorously with benzene. The formula $Ni(CN)_2.NH_3.C_6H_6$. The compound $Ni(CN)_2.NH_3.C_6H_6$ was originally thought to be co-ordination compound in which there were four groups bonded to the nickel, including the benzene molecule.

It has been shown that the compound has a very interesting structure in which half of nickel ions have a co-ordination number of six because ammonia molecules are co-ordinated above and below the plane. The six co-ordinate nickel ions are bonded to the nitrogen atom of the CN^- group. The benzene molecule is trapped in the 'holes' of the lattice because it just fits.

(ii) *Thiocyanate tetrakis* (4-methyl *pyridine*) *nickel* (*II*) *clathrates.* **Schaffer** (1957) has described many clathrates of this complex with benezene derivatives; they show a striking selectivity for para-isomers; pure meta xylene has been prepared on a semicommercial scale from a meta/para eutectic mixture by clathration of the para isomer.

This complex forms a mole per mole clathrate by simple contact with para-xylene. This fact indicates that this clathrate could be of the channel type, although its structure is not yet known.

(ii) *Thiocyanato tetrakis* (*alpha–phenyl alkylamine nickel* (*II*) *clathrates*). **Hanotier** (1965) has prepared more than seventy complexes of the type $[Ni(CNCS)_2(y - C_6H_4)CH(R)(NH_2)_4]$ in which R is a primary alkyl group and y can be a wide variety of substituents on any position of the benzene ring; it is almost always possible to select one of those complexes for clathrating selectively any desired isomer in a mixture of aromatic molecules. These clathrates have been used to bring about the separation of optically active amines.

6. Other Clathrates

(a) Several compounds have a monocylic structure having in its centre a hole which can accommodate foreign molecules, *e.g.*, tri–o–thymotide, cyclodextrins, desoxycholic acid, cycloveratryl and few others. Graphite gives sandwich type inclusion compounds with alkali metals and with fluorine.

(b) Jeffrey and Mak (1965) have described an interesting intermediate between true clathrates and hydrogen–bonded hydrates. In hexamethylene tetramine, hexahydrate, $(CH_2)_6N_4 6H_2O$, the hexamethylene tetramine molecule is trapped inside a cage composed of water molecules.

(c) Powell and Welters (1959) prepared a clathrate of composition $(C_{18}H_{20}O_2)_6$ Ar, which on heating liberates more than fifteen times its own volume of gas (STP). The host substance is condensation product of phenol and mesityl oxide, called 4– (p–hydroxyphenyl) 2, 2, 4– trimethylchroman.

(d) A similar clathrate $(C_7H_8O)_6$ Xe was prepared by **Trofimov** and **Kazankin** (1965) by subjecting *p*– cresol to 10 – 50 atm. pressure of xenon. The heats of dissociation of the clathrate into liquid p–cresol and gaseous xenon and into solid p–cresol and gaseous xenon are 96.6 and 35.6 kJ respectively.

APPLICATIONS OF CLATHRATES

1. Clathrates of Werner complexes have been used to bring about the resolution of optically active amines.
2. Several factories, especially in U.S.S.R, use clathrates of urea for dewaxing petroleum products and for producing straightchain paraffins.
3. Several processes based on the formation of hydrates mostly propane–hydrate clathrate have been used for the desalinization of sea water.
4. The radioactive Kr^{85}/quinol has been proposed as a safe and convenient source of beta radiation.
5. Radiometric determination of gaseous fluorine in concentration of only a few parts per million has been made possible by the reaction of the fluorine with quinol/Kr^{85}clathrate.
6. Noble gas clathrates are used for storing noble gases. This eliminates both pressure and volume problems.
7. Clathrate compounds can be utilized to study the physical properties of isolated molecules under conditions when

measurements could not normally be made.

8. Clathrate compounds have become important in the separation of hydrocarbons.
9. It is expected in the near future that noble gas clathrates might be used in handling and disposal of radioactive noble gases and as compact energy sources.
10. As helium and neon fail txo form clathrate compounds with hydroquinone whereas the Ar, Kr and Xe form such compounds, the former two inert gases can be separated from the latter three.
11. As the rare gas is immediately recoverable, either by heating or dissolution, the clathrates are used as temporary but convenient storage media in certain flow process.
12. Clathrate formation is of great use because it trapes the Kr^{85}, the radioactive isotope, produced by the fission of uranium in the nuclear reactors.
13. Kr^{85} having its long half–life of 10.3 years is found to be a convenient and compact source of energy for a small atomic battery, when used in the clathrate form.
14. The clathrate compounds can provide a very better way of examining megnetic susceptibilities of free molecules below the temperature at which they will normally condense to a liquid.
15. Clathrates are used to study the physical properties of isolated molecules under conditions when it becomes difficult to carry out the measurement normally.

COORDINATION COMPOUNDS WITH MONODENTATE LIGANDS

Sequence of central atom and ligand names : The ligands are listed in alphabetical order, without regard to charge, before the names of the central atom. Numerical prefixes indicating the number of ligands are not considered in determining that order.

1. Dichloro [diphenylphosphine (thiourea) platinum(II)]
2. Dibromobis [trimethylphosphine platinum(II)][1].

Number of ligands in a coordination entity : Two kinds of numerical prefix are available for indicating the number of each kind of ligand within the name of the coordination entity, etc. The simple di-, tri-, etc., derived from cardinal numerals, are generally recommended. The prefixes bis-, tris-, tetrakis-, derived from ordinals, are used with complex expressions and when required to avoid ambiguity; for example, one would use diammine but bis (methylamine) to make a distinction from dimethylamine.

When the latter multiplicative prefixes are used, enclosing marks are placed around the multiplicand. Enclosing marks are not required with the simpler prefixes di-, tri-, etc. There is no elision of vowels or use of a hyphen in tetraammine and similar names, except for compelling linguistic reasons.

Charge numbers, oxidation numbers, and ionic proportions: When the oxidation number of the central atom can be defined without ambiguity, it may be indicated by appending a roman numeral to the central atom name.[2] This number is enclosed in parentheses after the part of the name denoting the central atom. No positive sign is used. When necessary a negative sign is placed before the number. Arabic zero indicates the zero oxidation number. No space is left between this number and the rest of the name.

Alternatively, the charge on a coordination entity may be indicated. The net charge is written in arable numbers on the line, with the number preceding the charge sign, and enclosed in parentheses.

[1][The boldface italic letters are those used in the alphabetical placement of ligands names. Other, nondetermining letters are marked with "strike-throughs."]

[2][This is the Stock system of indicating oxidation state of the metal (and indirectly, charge on the complex). Stock, A.Z. Angew. chem. 1919, 27, 373.)

[3][This is the Ewens-Bassett system of indicating charge on the complex (and indirectly, the oxidation state of the metal). Ewens, R.V.G.; Bassett, H. Chem. Ind. 1949, 27, 131.]

Terminations for names of coordination entities : All anionic coordination entities take the ending-ate, whereas no distinguishing termination is used for cationic or neutral coordination entities.

Uses of Enclosing Marks

The formula for the entire coordination entity, whether charged or not, is enclosed in square brackets. When ligands are polyatomic, their formulae are enclosed in parentheses. Ligand abbreviations are also enclosed in parentheses. In the special case of coordination entities, the nesting order of enclosures is as given. There should be no space between representations of ionic species with in a coordination formula.

Examples:

1. $[Co(NH_3)_6]Cl_3$
2. $[CoCl(NH_3)]5Cl_2$
3. $[CoCl(NO_2)(NH_3)_4]Cl$
4. $[PtCl(NH_2CH_3)(NH_3)_2]Cl$
5. $[K_2[PdCl4]$
6. $[Co(en)_3]Cl_3$.

Ionic charges and oxidation numbers : If the formula of a charged coordination entity is to be written without that of the counterion, the charge is indicated outside the square bracket as a right superscript, with the number before the sign. The oxidation number of a central atom may be represented by a roman numeral used as a right superscript on the element symbol.

Examples:

1. $[PtCl_6]^{2-}$
2. $[Cr(H_2O)_6]^{3+}$
3. $[Cr^{III}(NCS)_4(NH_3)_2]^-$.

COORDINATION CHEMISTRY CONCEPT

The compounds contain a central atom or ion. Usually a metal, which is chemically bonded to various groups is called the *acceptor* and the attached groups are known as *donor groups or ligands.* For example, in the complex ion $[Fe(CN)_6]^{4-}$,Fe^{2+} ion is the central metal ion or acceptor while CN^- ions are ligands.

LIGAND

Ligands or Coordinating Groups and Central Metallic Atom

The neutral molecules or ions (usually anions) which are attached with the central ion in complex compounds are called ligands or co-ordinating groups. For example in the complex ion, $[Fe(CN_6)]^{3-}$ the six CN~ ions which are attached with the central Fe^{3+} ion as shown in the margin act as ligands.

In Lewis sense, in most of the complex compounds the ligands act as Lewis bases (electron pair donors) and the central metal ion acts as a Lewis acid (electron pair acceptor), *i.e.*, in most of the complex compounds the ligands donate one or more electron pairs to the central metal ion.

M^{n+}	+	x L	→	$[ML_x]^{n+}$
Central metal ion(Lewis acid or electron pair acceptor)		Ligand (Lewis base or electron pair donor)		Complex compound

In some complex compounds the ligand acts both as donor and acceptor. For example, in metallic carbonyls, CO molecules which act as ligands act both as donor and acceptor ($M \rightleftharpoons CO$). In a ligand the atom which actually donates the electron pair to the central metal ion is called *donor* or *coordinating atom*. The ligands are attached with the central metal ion through their donor atom (or atoms). The metallic atom with which the ligands are attached through coordinating bonds is called central metallic atom. This metallic atom may be in zero, positive or negative oxidation state.

Coordination Number of the Central Metal Atom/ion

Coordination number of the central metal atom/ion in a given complex compound is equal to the total number of donor atoms which are actually attached with the central metallic atom. In other words we can say that the coordination number of the central metallic atom is equal to the number of sites at which the ligand(s) is attached with the central metallic atom.

In case of complex compounds which contain only monodentate ligands, the coordination number of the central metallic atom is equal to the number of monodentate ligands coordinated to the metallic atom. This rule does not hold good for the complexes containing polydentate (*i.e.*, bidentate, tridentate, tetradentate etc.) ligands. Coordination number of the metallic atom predicts the geometry of the complex compound. Thus, for coordination number equal to 2, 3, 4, 5 and 6, the geometry of the complex compound is *linear, trigonal planar, tetrahedral (or square planar), trigonal bipyramidal and octahedral* respectively. This discussion shows that the coordination number gives us an idea about the way in which the ligands are arranged round the central metallic atom.

Coordination Sphere and Ionisation Sphere

While writting the structural formula of a given complex compound the central metal atom and the ligands attached with it are always written in a square bracket, [ ].

This square bracket is called *coordination* (or *inner*) *sphere.* The portion outside the coordination sphere is called ionisation (or outer) sphere. Thus in $[Co(NH_3)_5Cl]Cl_2$, the square bracket which contains the central metal ion (Co^{3+} ion) and the ligands (five NH_3 molecules and one Cl^- ion) is coordination sphere and the portion that contains two CF ions is ionisation sphere.

The species written in ionisation sphere are ionisable and hence can be precipitated by means of a suitable precipitating agent while those given in the coordination sphere (*i.e.* metallic atom and ligands) are non-ionisable and hence cannot be precipitated. This is shown below for

$$[Co(NH_3)_5Cl]Cl_2 \rightleftharpoons [Co(NH_3)_5Cl]^{2+} + 2Cl^-$$

Two Cl^- ions present in ionisation sphere can be precipitated as AgCl (white ppt) by adding.

$$2Ag^+ + 2Cl^- \rightarrow 2AgCl \downarrow$$

Cl^- ion written in coordination sphere is not ionisable and can, therefore, not be precipated.

CLASSIFICATION OF LIGANDS

Depending on the number of sites at which one molecule of a ligand is coordinated to the central metallic atom, the ligands have been classified as monodentate (or unidentate) and polydentate (or multidentate) ligands.

1. *Monodentate or unidentate ligands :* The ligands which have only one donor atom or are co-ordinated through one electron pair are called monodentate or unidentate ligands. Such ligands are coordinated to the central metal ion at one site or by one metal-ligand bond only. These ligands may be neutral molecules or in anionic form.

2. *Polydentate or multidentate ligands :* These may be *bidentate, tridentate, tetradentate, pentadentate* and *hexadentate*, if the number of donor atoms present in one molecule of the ligand attached with the central metallic atom is 2, 3, 4, 5 and 6 respectively. Thus one molecule of these ligands is coordinated

to the central metallic atom at 2, 3, 4, 5 and 6 sites respectively. In other words, we can say that one molecule of these ligands makes 2, 3,4, 5 and 6 metalùligand coordinate bonds respectively.

Any atom, ion or molecule which is capable of donating a pair of electrons to the metal atom is called a co-ordinating group or ligand. In a ligand, the particular atom which actually donates the electron pair is called the donor atom. For example, in the complex potassium ferrocyanide $K_4[Fe(CN)_6]$, the six $(CN)^-$ ions are ligands and the nitrogen in $(CN)^-$ is the donor atom. Ligands may be classified as unidentate and polydentate ligands.

The ligands have been found to be arranged around the central metal ion inside the first sphere of attraction in preferred geometries. The common geometries found in complexes are linear, equilateral, triangular, tetrahedral, square planar, trigonal bipyramidal, square pyramidal and octahedral.

TYPES OF LIGANDS

Ligands have been classified in two ways :

1. Classification based on donor and acceptor properties of the ligands : Such ligands have been further classified as follows :

(a) *Ligands having one (or more) lone pair (or pairs) of electrons.* These ligands are of the following two types :

(i) *First type includes such ligands which have vacant* p-*type orbital that can receive back donated* p-*electrons from the metal ion in low oxidation state.*

The main examples of such ligands are CO,CN, isocyanides, NO, R_3P, R_3As, a, a-*dipyridyl, o-phenanthroline and unsaturated organic molecules. During the formation of complexes, these ligands as well as metal atoms act both as donors and acceptors* $(M \underset{\pi}{\overset{\sigma}{\rightleftarrows}} L)$. *The reason is that these ligands have filled donor orbitals in addition to vacant* p-*type acceptor orbitals.*

(ii) *Second type includes such ligands which have no vacant orbitals to receive back donated electrons from the metal. Examples of this type are* H_2O, NH_3, F^- *etc.*

(b) *Ligands having no lone-pairs of electrons but having* p- *bonding electrons.* Examples of these ligands are ethylene, benzene, cyclopentadienyl ion, etc.

2. Classification based on the number of donor atoms present in the ligands. These ligands have been classified as follows :

Monodentate ligands : If a ligand contains only one atom *i.e.* it is capable of forming only one coordinate bond to the central metal atom, it is known as *monodentate or unidentate ligand.*

The monodentate ligand may be any of the following types :

(i) Monodentate ligands may be those in which the ligand atom only contains a π-bonding lone pair, *i.e.*, it does not contain π- electrons and no vacant orbitals.

Such ligands are generally good Lewis bases. They belong to the first short period of the periodic table. Examples of these are H–, NH_3, SO_3^{2-} and neutral aliphatic amines.

(ii) Monodentate ligands may include those in which the ligand atom contains three lone electron pairs. These split under the influence of bonding to metal atom into two π-orbitals of higher energy and one (3-orbital of lower energy. Examples of these are N3⁻, O^{2-}, OH^- S^{2-}, CT^- Se^{2-}, Br^-, I^- etc.

(iii) Monodentate ligands may include those in which the ligands atom contains two lone pairs. One of these is used for σ-bonding while the other is used for π-bonding. Examples are , R_2O NH_2^- R_2S, R_2O etc.

(iv) Monodentate ligands may include those in which the ligand contains a σ-bonding pair along with low-lying empty π-antibonding orbitals which can accommodate back donation from metal to ligand (M ⟶ L).

(v) Monodentate ligands may include those in which the ligand atom has no unshared electron pairs but contains electrons already involved in intramolecular π-bonding. Examples are alkenes, alkynes, benzene, etc.